KS3 Science

Book 3

Collins

£1-40

Ed Walsh
Series Editor

Tim Greenway
Ray Oliver
David Taylor

Published by Collins
An imprint of HarperCollinsPublishers
77–85 Fulham Palace Road
Hammersmith
London
W6 8JB

Browse the complete Collins catalogue at
www.collinseducation.com

© HarperCollinsPublishers Limited 2008

10 9 8 7

ISBN 978-0-00-726422-3

British Library Cataloguing in Publication Data. A Catalogue record for this publication is available from the British Library.

Commissioned by Penny Fowler
Project managed by Laura Deacon
Glossary written by Pam Large
Edited and proofread by Camilla Behrens,
Anita Clark, Rosie Parrish and Lynn Watkins
Internal design by Jordan Publishing Design
Page layout by eMC Design Ltd, www.emcdesign.org.uk
Illustrations by Jorge Santillan, Peters & Zabransky

Production by Arjen Jansen
Printed and bound by Printing Express, Hong Kong

MIX
Paper from
responsible sources
FSC™ C007454

FSC™ is a non-profit international organisation established to promote the responsible management of the world's forests. Products carrying the FSC label are independently certified to assure consumers that they come from forests that are managed to meet the social, economic and ecological needs of present and future generations, and other controlled sources.

Find out more about HarperCollins and the environment at
www.harpercollins.co.uk/green

Contents

Introduction

Exciting Topic Openers

Every topic begins with a fascinating and engaging article to introduce the **topic**. You can go through the questions to see how much you already know – and you and your teacher can use your answers to see what level you are at, at the start of the topic. You can also see what the big ideas are that you will cover in this topic.

Levelled Lessons

The colour coded levels at the side of the page increase as you progress through the lesson – so you can always see how you are learning new things, gaining new skills and **boosting** your level. Throughout the book, look out for fascinating facts and for hints on how to avoid making common mistakes. The keywords along the bottom can be looked up in the glossary at the back of the book.

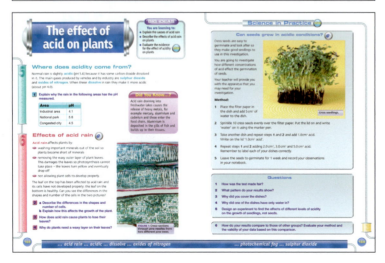

Get Practical

Practicals are what Science is all about! These lessons give you instructions on how to carry out your experiments or investigations, as well as learning about the Science behind it.
Look out for the HSW icons throughout all of the lessons – this is where you are learning about How Science Works.

Welcome to Collins KS3 Science!

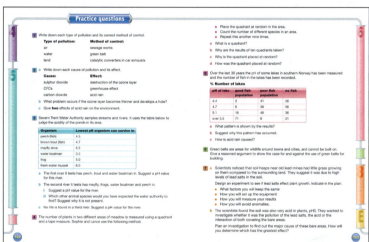

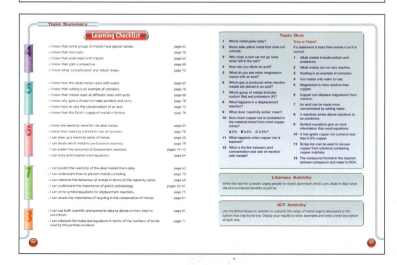

Mid-Topic Assessments

About half way through the Topic you get a chance to see how much you've learned already about the big ideas by answering questions on a stimulating article. The Level Booster allows you to see for yourself what level your answers will reach – and what more you would need to add to your answers to go up to the next level. There are also opportunities here to see how the Topic relates to other subjects you are studying.

Practice Questions

At the end of the Topic you can use these questions to test what you have learned. The questions are all levelled so you can stretch yourself as far as possible!

Topic Summary

Check what you have learned in the levelled Learning checklist - you can then see at a glance what you might need to go back and revise again. There's a fun Quiz and an activity linked in to another subject so you can be really sure you've got the Topic covered before you move on to the next one.

How much alcohol is there in each drink?

On the unit system an adult drinking more than five units of alcohol would be over the limit to drive. This would give a blood alcohol content of 80 mg/litre. The unit system also recommends that adults do not have more than 21 units in a week or their alcohol consumption may start to cause serious long-term effects. That means a maximum of three to four units a day for men and two to three units for women, with at least one alcohol-free day per week. The 21 units should not be drunk in a binge at the weekend.

One unit is the same as $10\,cm^3$ of alcohol, regardless of how diluted. The unit system is misleading though. One of the problems is that people are different sizes and so the same amount of alcohol may cause a smaller person to be affected more. Secondly, the alcohol content of drinks can vary, depending on the strength of the drink. The unit system was based on a certain strength of each drink. For beer and cider it was worked out for a 4% alcohol content and for wine at an 8% alcohol content. The table below gives the range of some of the more common drinks. Homemade brews of beer, wine and cider can be far stronger.

FIGURE 1: Half a pint of beer is one unit.

FIGURE 2: A small tumbler of spirits is one unit.

FIGURE 3: A glass of wine is one unit.

FIGURE 4: Half a pint of cider is one unit.

One pint of lager (4%) would be two units, one for each half pint, whereas one pint of strong lager (6%) would be three units.

One standard glass of wine (12%) would be two units not one.

One pint of strong cider (6%) would be three.

There is a need to inform the public of the real alcohol content of the drinks they consume. A person may drink at lunch time and then later in the day and find that they are over the limit because the alcohol has not been removed from his or her body. The liver can only remove one unit every hour, there is no way of making the alcohol disappear faster.

BIG IDEAS

By the end of this unit you will be able to describe some of the factors that affect human behaviour and explain ideas about learning. You will understand how some drugs can have a damaging effect on the way in which the human body operates, and you will have communicated scientific ideas in a variety of ways.

What do you know?

1 Give **two** examples of drinks that are called spirits.

2 What fruit is used to make **a** wine **b** cider?

3 Why is it dangerous to drink and drive?

4 What happens if you are caught drinking and driving?

5 Which drinks have the most alcohol in?

6 Should it be made illegal to drink and drive?

7 At what age can you legally buy alcohol in a pub?

8 In certain states in America the legal age to drink alcohol is 21 years. Should the legal age be raised or lowered in Britain?

9 If I drank four halves of strong beer why might I be over the limit for driving?

10 Why is it inadvisable to use the unit system if you are going to drive?

11 Design a way of informing people of the amount of alcohol in their drink.

Drink	% alcohol content	Drink	% alcohol content
Beer	4–8	Wine	8–16
Lager	3–6	Sherry	16–18
Cider	4–8	Gin/whisky/vodka	35–50

What is a drug?

BIG IDEAS

You are learning to:
- Explain the problems caused by addiction
- Explain the difference between stimulants and depressants
- Evaluate the problems caused by drug taking

Homeless

The streets of many cities have people who sleep in doorways, on benches and in cardboard boxes. Often, these people are young and look lost. It is very easy to try to ignore them and to label them.

What are drugs?

If a person is sleeping rough, they may be homeless and they may end up taking illegal drugs. Drugs alter our behaviour by affecting the brain. They may relax a person, remove pain or make them feel more energetic. Some drugs are legal such as caffeine and alcohol, some are prescribed such as medicines and laws restrict the use of other drugs.

1. Describe what you think the picture shows.
2. Suggest why the person may have ended up like this.
3. Name **one** legal drug people might take regularly.
4. Name **one** drug given as a medicine.

FIGURE 1: How do people end up homeless?

Types of drugs

Drugs can be put into the following broad groups.

Stimulants: which increase brain activity, e.g. caffeine, cocaine.

Depressants (or sedatives): which slow down the nervous system e.g. alcohol, heroin, and tranquillisers.

Pain-killers e.g. paracetamol

Hallucinogens: which cause the mind to 'see things' e.g. LSD

5. Describe the appearance of the tablets in Figure 2. What do you think they are?
   ecstasy   aspirin   worm tablet

6. Why might a person drink coffee when they have to stay up through the night?

7. Why might a person take tranquillisers if they cannot sleep?

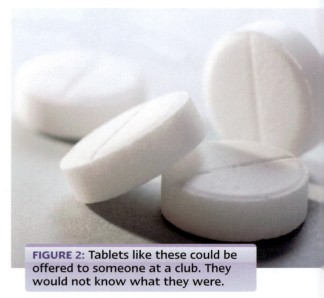

FIGURE 2: Tablets like these could be offered to someone at a club. They would not know what they were.

... addicted ... depressants ... stimulants

Addiction and withdrawal

When a person takes a drug their body may start to crave for it and they will become **addicted**. This means that they will suffer **withdrawal** symptoms when they try to stop taking the drug.

Coming off heroin can be a shattering experience. Withdrawal symptoms are:

- after about four hours the person is restless and anxious
- after eight hours they sweat and fluids are discharged from the nose and eyes
- after around 16 hours the pupils dilate, they get hot and cold flushes, they tremble and their muscles and bones ache
- after around 20 hours an inability to sleep, severe restlessness, fever and a feeling of sickness occur. Heart rate is high.

This is then followed by vomiting, abdominal pain and diarrhoea. After 6–7 days the person is left weak and exhausted. Any dose of heroin causes addiction.

FIGURE 3: Why do withdrawal symptoms happen?

8 Explain why it is difficult for drug addicts to come off a drug.

9 Explain why there is a link between drug addiction and crime.

Overdose

A heroin user finds they have to increase the dose to have the same effect. The body has developed **tolerance** to the drug. This effect can occur with several other drugs. The drug addict may then start using high amounts to stop the withdrawal symptoms and this gets to a level where they have too much for their body to cope with. This is when they are likely to take an overdose.

10 Why does a drug addict have to take more and more of a drug to get the same effect?

11 High levels of caffeine cause addiction and withdrawal effects.
   a What are the effects of caffeine?
   b Explain whether you think children under 16 should be allowed coffee.

12 Methadone is a drug used to help heroin addicts come off the drug. It is still addictive. Research the effect of methadone. Explain why it is used and why it is not always successful.

Did You Know...?

A person can become addicted to dangerous events such as extreme sports. The brain produces chemicals called endorphins, which give the person a 'buzz'.

Is alcohol really that good?

BIG IDEAS

You are learning to:
- Explain the effect of alcohol on the body
- Evaluate the danger of alcohol
- Give reasoned arguments for and against alcohol being made illegal

Is alcohol a dangerous drug?

Alcohol is a drug. It is a depressant. This means it slows down the rate that a person's body reacts. In the UK, 30% of injuries to pedestrian victims of road traffic accidents and more than half of those to drivers are linked to alcohol consumption.

So is alcohol a dangerous drug?

Yes

Alcohol is addictive with severe withdrawal symptoms.

Alcohol is a poison that can cause long-term damage to the liver, heart, brain, sex organs and the kidneys. Cancers can also occur.

Alcoholics often suffer from muscle wasting and lose bone mass.

Drinkers are more likely to have casual sex and that can lead to STDs.

One in 10 men and one in 20 women have an alcohol problem.

No

The liver can remove alcohol at the rate of one unit per hour.

Research has shown a moderate amount of beer and red wine is good for the heart.

Moderate alcohol levels in young women reduce blood pressure.

Drinking helps people relax socially.

> **Did You Know...?**
>
> When you consume alcohol you lose more water in your urine than you drink.
> After a heavy night drinking your body is dehydrated, which causes the brain to shrink away from the skull. This is what causes a hangover.

> **Did You Know...?**
>
> Alcohol causes addiction and tolerance. When an alcoholic stops drinking alcohol he or she will go through withdrawal and will then need rehabilitation. The symptoms of withdrawal are sweating, insomnia, hallucinations, cramps, vomiting and fever. This may go on for more than three days.

1. What effect does a depressant drug have on the user?
2. Why does alcohol increase the risk of accidents?
3. What is meant by the term alcoholic?
4. Explain whether you think alcohol is a dangerous drug. Give clear reasons for your answer.

... alcoholics ... alcohol poisoning ... blood alcohol concentration (BAC)

Effect of dosages

The effects of alcohol on the body are shown below matched to the **blood alcohol concentration (BAC)**. This varies from individual to individual due to factors like tolerance level and size.

Feeling good (BAC 0.03–0.10%)

- Person more self-confident and daring.
- Attention span shorter.
- Judgement poor. Talks too much, saying the first thing that comes into their head.

Tiredness (BAC 0.06–0.25%)

- Person becomes sleepy.
- They have trouble understanding or remembering what happens.
- Body movement uncoordinated. Not able to walk properly.
- Vision blurs. Senses poor.

Confusion (BAC 0.11–0.30%)

- Very confused. Dizziness and staggering.
- Very emotional. Aggressive, withdrawn or over-affectionate.

- Poor **coordination**. Nausea and vomiting may occur.
- Do not feel much pain.

Stupor (BAC 0.21–0.40%)

- Movements impaired. Lapse in and out of consciousness.
- Person may go into a **coma**.
- Risk of death very high due to **alcohol poisoning**.
- Loss of bodily functions may begin.

Coma (BAC 0.30–0.50%)

- Unconsciousness occurs.
- **Reflexes** not responding.
- Breathing and heart rate far slower.

Death (BAC more than 0.40%)

- Nervous system fails causing death.

Questions

1   Explain why a person may have several bruises and cuts from falling after drinking but does not remember it.

2   Explain why you cannot predict the exact effect of drinking alcohol.

3   The liver breaks down alcohol at a set rate. Chronic alcoholics overload the pathway so the liver cells are destroyed. Explain the effect on the body if the liver cells are destroyed.

4   Use your ideas on the effect of alcohol on the body to present a case for or against making alcohol an illegal drug.

5   One unit of alcohol has a BAC of 0.016%. Present the information above in an appropriate form that shows the main headings and the number of alcoholic drinks consumed. The drinks to consider are a normal beer/cider, a strong beer/cider, a large glass of wine, and a spirit. The graph can be a pie chart, a block graph or even a table.

A nail in the coffin

BIG IDEAS

You are learning to:
- Describe the contents of a cigarette
- Explain how smoking affects the body
- Evaluate models of the effects of smoking

A dangerous habit

Smoking was very fashionable in the 1940s as film stars were regularly shown smoking. It was then found a lot of smokers were becoming ill.

When you smoke a cigarette the main chemicals it produces are:

tar   nicotine   carbon monoxide.

Low tar cigarettes contain a reduced amount of tar, but the levels are still very high. The amount of nicotine in cigarettes has also been reduced. It was far higher in the 1950s, so it made even more people want to smoke. Nicotine is the drug that causes addiction.

Carbon monoxide joins to red blood cells and stops them carrying oxygen.

Smoking also produces acid gases, which increase the risk of tooth decay.

1. **What reasons are given for people smoking today?**
2. **What illnesses do you know that are caused by smoking?**
3. **Which chemical in the cigarette causes the addiction?**
4. **Why might more people have been hooked on smoking in the 1950s?**

FIGURE 1: Cigarettes

Did You Know...?

Pipe smoking was popular amongst British explorers in Africa as the fumes kept away many of the mosquitoes. Unfortunately, pipe smoking causes the same effect as cigarettes.

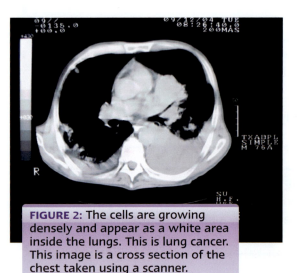

FIGURE 2: The cells are growing densely and appear as a white area inside the lungs. This is lung cancer. This image is a cross section of the chest taken using a scanner.

How the lungs suffer

When a person smokes, tar becomes trapped in the mucus in the lungs. It starts to paralyse the cilia. Remember the cilia are the hairs on the cells lining the bronchioles. They sweep the mucus out of the lungs.

The tar will then irritate the cells lining the bronchioles and cause them to become inflamed, causing **bronchitis**. Further damage may occur, with the possibility of **cancer** as the cells grow abnormally.

... bronchitis ... cancer ... emphysema

The person will also cough a lot and this can cause damage to the alveoli. The damage to the alveoli means less oxygen can be taken in through the alveoli, so the person becomes breathless resulting in **emphysema**. The person will then need to be provided with oxygen to breathe. Some of the carbon monoxide inhaled is absorbed by the red blood cells so less oxygen is carried by the blood.

5 Why do smokers cough, especially in the morning?

6 Why might a smoker feel breathless?

How Science Works

Smoking Machine

The following apparatus (which has to be used in a fume cabinet) shows the effect of smoking on the body.

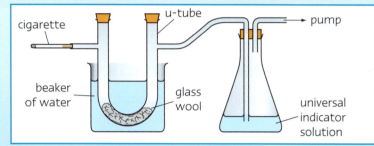

cigarette — u-tube — pump
beaker of water — glass wool — universal indicator solution

1 a Which part acts as the lung?
   b What results are shown when one cigarette is smoked?

2 Explain, using the evidence, why smoking increases the risk of tooth decay.

Can it really cause the loss of a leg?

Another serious effect of smoking is to increase the smoker's heart rate, as less oxygen is carried in the blood. Smoking increases the **risk** of blockage in arteries as the blood vessel walls become less elastic. It also causes narrowing of the arteries supplying the limbs.

If this narrowing occurs in the blood vessels to the heart muscles, less oxygen will reach the muscles. If it occurs in blood vessels to the limbs then the cells start to die off.

It can also damage arteries to the brain resulting in strokes.

How Science Works

A nicotine patch works by releasing small amounts of nicotine to be absorbed by the skin. This is meant to reduce the craving for a cigarette.

FIGURE 3: Do you think the fashion to smoke is changing?

Circulation problem	Increase risk of a smoker over a non-smoker
Heart attack	2–3 times
Blocked blood vessels in legs or arms	16 times
Stroke	2–4 times

7 Explain how smoking reduces an athlete's performance.

8 Suggest how the information of the increased risk of a smoker to a non-smoker was obtained.

9 Explain how and why a) lung cancer and b) emphysema affect the rate of respiration.

Another nail in the coffin

Risk levels for smokers

A person who smokes increases their **chance** of dying from smoking-related diseases.

Facts

- On average, smokers die 10–15 years earlier than non-smokers.
- 25% of teenagers who smoke die from smoking-related diseases.
- Over 80% of smokers start as teenagers.
- The highest rate of smoking is in North East England.

How Science Works

Number of cigarettes smoked per day	Annual death rate per 100,000 people from smoking-related diseases
0	10
1–14	78
15–25	127
25 or more	251

1. Plot a graph of the number of cigarettes smoked and the increase in risk of dying from smoke-related diseases. (You will need to work out the risk by comparing the death rate of the smoker to the non smoker.)

2. If I smoke 20 cigarettes a day how much have I increased my risk of dying from smoke-related diseases?

3. What pattern is shown by the data?

FIGURE 1: Does it matter if someone smokes near you?

What about non-smokers?

Second-hand smoking or passive smoking is a major source of indoor air pollution. The short-term effects can be coughing, headaches, sore throats, breathing problems and heart problems.

The long-term effects are risk of heart problems, respiratory disease and lung cancer.

Facts

- Non-smokers who are exposed to smoking at home have a 25% increased risk of heart disease than normal.
- The domestic exposure to second-hand smoke has caused 2700 deaths a year in people aged 20–64.
- Exposure to second-hand smoke at work causes 617 deaths per year.

4 Explain why a second-hand smoker has a cough and breathing problems.

5 Explain why the government has banned smoking in public places.

How come not everyone gets lung cancer?

Every person is different. They are a result of their genetic information being influenced by the environment. A person may have information in their genes that makes them more likely to develop cancer if they are in certain environments. Therefore if they smoke they are more than likely to get cancer.

6 Explain why one person who smokes regularly lives until they are 60, whereas another person dies at 30 from lung cancer.

7 Explain, giving clear reasons, whether you think a patient should receive a heart or lung transplant if it can be shown that their condition was caused by their smoking.

8 Early experiments on the effects of smoking were carried out on dogs. What are the advantages and problems caused by using animals to investigate the effects of smoking on humans?

FIGURE 2: If you smoke are you taking a gamble with your life?

Did You Know...?

Every year 114 000 smokers die from smoking-related diseases. Around 10 million people smoke in the UK.
The highest rate is amongst the 20–24 year olds.

Cannabis

BIG IDEAS

You are learning to:
- Explain the effects of cannabis on the body
- Evaluate the evidence for and against the use of cannabis
- Interpret information on the effects of cannabis

Cannabis

Cannabis is made from the dried flowers and leaves of the cannabis plant. **Hashish** is formed from the leaves and stems. The cannabis plant is related to nettles. It has been used for medical purposes for thousands of years.

Cannabis can also be called **marijuana**, ganja, hashish, pot, weed or grass. It is an illegal drug used for leisure by some people. Doctors in hospitals can also prescribe it for medical use. It makes a person feel relaxed and dreamy. The most common way of taking cannabis is smoking, but it can also be eaten.

FIGURE 1: Cannabis leaves.

Smoking cannabis

Cannabis contains more tar than tobacco plants, and a higher concentration of carcinogens (cancer causing chemicals). If it is smoked it can cause bronchitis, emphysema and lung cancer just like cigarettes. Cannabis is less addictive than cigarettes and there are few signs of withdrawal. The effects on users are cheerfulness, relaxation and light-headedness. The person may also have mild hallucinations.

1 Which part of the cannabis causes lung cancer?

2 What addictive drug found in cigarettes is not found in cannabis?

3 Compare the similarities in the health effects of smoking cannabis and cigarettes.

FIGURE 2: Why is this cannabis being grown in a room with no windows?

... cannabis ... hashish

Should cannabis be legalised?

Many studies have been done on the effects of cannabis. One was a study of 129 college students who smoked marijuana. Some of the results are given below.

- Those who smoked marijuana regularly in the 30 days before a test showed poor memory, low attention span and poor learning compared with those not using marijuana, even if they had not taken marijuana for 24 hours.
- Heavy marijuana abusers could not sustain their attention, nor organise their thoughts.
- The ability of heavy marijuana users to recall words from a list improved after quitting for over four weeks. One week made little difference.

Case for legalisation	Case against legalisation
It is beneficial as a painkiller for cancer, victims multiple sclerosis and AIDS victims.	It is slightly addictive and heavy users can display aggression.
Reports indicate it has no strong link to harder drug use.	It can cause mild panic and paranoia. It can also result in short-term memory loss.
It can affect the heart in similar ways to exercise.	Cannabis may be the 'gateway' to trying more dangerous drugs.
It causes relaxation.	Cannabis can have lasting effects on the memory, affecting the ability to learn.
It does not cause as many health problems as alcohol.	It has the same effects as smoking.

4 Use the evidence to explain how marijuana affects a person's capability.

5 Use your ideas to discuss how useful the evidence from the study is.

6 Give a reasoned case for or against cannabis being a legal drug.

What the law says

Cannabis has been already been downgraded from Class A to Class C. Drugs in Class A have a greater penalty for possession, up to seven years. Production, possession and distribution of cannabis is however still illegal. If you are caught with cannabis you could receive up to two years in prison, whereas supplying cannabis has a sentence of 14 years.

7 Why is supply given a higher sentence than possession?

8 Suggest why the government might have downgraded cannabis.

9 What evidence would you need to collect to find out if:
   a cannabis leads to people trying more dangerous drugs;
   b cannabis has long-term effects on the memory.
   How would you use this evidence?

Did You Know...?

44% of 16–29 year olds have reported to have tried cannabis at some point in their lives. 76% of people arrested in 1998 were charged with possession of cannabis.

Just say no

BIG IDEAS

You are learning to:
- Describe the effects of different types of drug
- Explain the dangers of taking each of the drugs
- Evaluate the dangers of different drugs

Why take drugs?

Headings in papers indicate over 50% of teenagers have tried cannabis. Is that number right? Would teenagers want to admit that they had taken cannabis? Or would they say they had tried it even if they had not?

Why do people take drugs? Evidence shows many reasons:

- peer pressure
- experimenting
- curiosity
- wanting to try it because it is not allowed
- thinking it will help deal with life's difficulties.

1 Do the papers put pressure on young people to take drugs or not take drugs?

2 What is the meaning of peer pressure?

3 How can you overcome peer pressure?

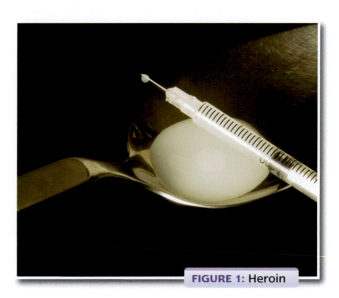

FIGURE 1: Heroin

FIGURE 2: Cocaine

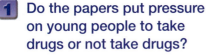

FIGURE 3: Ecstasy

... cocaine ... ecstasy ... heroin

Different types of drugs

There are many types of drugs; this table summarises some of the main drugs.

Drug	Short-term effects	Long-term effects
Heroin (from poppies)	Depressant causing a feeling of relaxation, warmth and lack of pain.	Highly addictive, causes tolerance. Withdrawal symptoms are severe and last 7–10 days. The addict shows neglect and poor nutrition.
Cocaine (crack cocaine)	Stimulant, causes sweating, loss of appetite and increased pulse rate. Snorting cocaine can damage the nose membranes.	Crack is highly addictive. Everyday use can cause restlessness, insomnia, weight loss and paranoia.
LSD (acid)	Hallucinogenic. Users report distorted shapes, intensified colours.	It is not addictive but the hallucinogens are very dangerous. People have died because after they have taken it, they think they have supernatural powers like being able to fly.
Ecstasy	Stimulant. The user experiences energy without anger. In extreme cases just one tablet can cause death.	Regular use can lead to sleep problems, lack of energy and depression. Short-term memory loss and emotional changes may occur.

4 Which of the drugs are stimulants?

5 Which of the drugs are addictive?

6 Why do heroin addicts find it so difficult to give it up?

Giving up drugs

The papers report cases of pop stars and film stars going to clinics to rehabilitate. This means they must go through withdrawal and then get their health back through improving their eating and dealing with the cause of drug dependence.

However, it is very difficult to cure yourself of a drug addiction and some people struggle for the rest of their lives.

7 Suggest why celebrities may start to take drugs.

8 Many ex-drug takers tour schools to talk about the dangers of drugs. Explain whether this is useful or not.

9 Explain, using your knowledge and understanding of drugs, which drug is the worst drug.

10 Explain why ecstasy can kill some people who take it.

How Science Works

Drug use can be detected by urine and blood samples. Hospitals may shine a light into a person's eyes to see if his or her pupils constrict; if they do not, the person may be high on drugs. HSW

Did You Know...?

Solvents such as glue, petrol, nail varnish and aerosols are dangerous drugs. They cause damage to the lungs and can cause heart failure. Long-term use causes brain damage, weight loss and liver and kidney damage.

Cycling's most famous race may have ended in Paris, but French police investigations into widespread drug taking have rocked the sport.

Although drug allegations are not new to cycling, the reputation of the Tour and the sport as a whole has been badly damaged and far reaching reform is expected.

The scandals began on the 15th July 1998. First a car belonging to the Festina cycling team was found to contain huge quantities of performance-enhancing drugs, including erythropoietin, which increases red blood cell counts so cyclists can increase their endurance. Then the team director admitted that some of his cyclists were routinely given performance-enhancing drugs.

The Festina team were expelled from the race, and soon French police and sports officials were raiding other teams' headquarters. On the 28th July four teams withdrew from the race and a further two dropped out on the 30th July.

Can they win only with drugs?

The Tour de France is the biggest, hardest and most gruelling cycling race in the world, a prize so precious people will do anything to win. In the 1960s riders attempted to get an edge by taking stimulants, and a rider lost his life doing so in 1967.

Cycling is not the only sport where competitors are willing to take drugs to win. Boxers have been found to use diuretics, which help the body lose fluid and as a result weight, this helps the boxer meet the weight limit for fighting.

The 1988 Seoul Olympics were tarnished by the disqualification of the 100 m gold medal winner Ben Johnson. He broke the world record by running 9.79 seconds, but was found to have used an anabolic steroid called stanozolol. He also showed signs of drug taking by the increase in his muscle bulk and the jaundiced appearance of his eyes. It is now thought Johnson had been made the scapegoat for far bigger problems of drug taking occurring in 1988.

A recent study of women body builders has shown widespread abuse of anabolic steroids. Anabolic steroids help the retention of protein, which aids in the development of muscle. This will result in the person's muscles increasing in mass and strength. But there are serious side effects.

In men:
- shrinking testes, reduced sperm count and impotence
- pain when urinating
- development of breasts.

In women:
- facial hair growth
- deepened voice
- menstrual cycle changes.

In both men and women:
- rapid weight gain
- clotting disorders
- liver damage
- heart attacks and strokes.

Drug testing has become more sophisticated to detect the new drugs that are being developed. Urine or blood samples are taken and analysed using:
- Gas chromatography and mass spectrometry. In each case the drug has a 'fingerprint' to match it to.
- Immuno-assays to identify the proteins (found in certain drugs) using antibodies tagged with dyes.

Assess Yourself

1 What samples are taken from athletes to test for drugs?

2 What tests are carried out on the samples?

3 What are the short-term benefits of athletes taking anabolic steroids?

4 What are the long-term problems of taking anabolic steroids?

5 Suggest why more cases of drug abuse in sport are being reported now than in the 1950s.

6 If an athlete is found to be taking drugs they are banned for two years, in Britain they will also be banned from representing their country at the Olympics. Is this too great a penalty or too lenient?

7 At present, an athlete must be able to provide a urine sample when tested at random throughout the year, if they miss three tests then they can be banned. This led to a famous footballer being banned for failing to give a sample. Is random sampling a good way to catch drug takers?

8 Concern has been expressed about whether the London Olympics will be drug free. Explain what the problems are in detecting drug users.

9 Outline the advances made in the detection of drugs.

Geography Activity

Find out more about the Tour de France. What is it about this particular event that makes it exceptionally strenuous?

PSHE Activity

Discuss whether any drugs should be used in sport. Should sports people who have prescribed drugs (e.g. asthma sufferers, people with cancer) be allowed to compete whilst taking drugs?

Level Booster

8 Your answers show an understanding of the developments in drug testing in sporting events and the problems that need to be overcome.

7 Your answers show an understanding of the problems faced by sporting bodies in controlling the use of drugs.

6 Your answers show you can give a detailed explanation of how a drug increases performance by affecting the body's metabolism.

5 Your answers show an understanding of how drugs can be detected.

Detecting your environment

BIG IDEAS
You are learning to:
- Describe the nerve pathway
- Explain the role of sense organs
- Explain why the nervous system may not function properly

Sense organs

A human has five **senses** and for each sense there is a **sense organ** that detects a **stimulus**. The sense organs and their stimuli are shown in the table below.

1. Which senses are likely to be well developed in a blind person?

2. Suggest why it is difficult for a person to taste food when they have a cold.

Sense	Sense organ	Stimulus
sight	eye	light
hearing	ear	sound
taste	tongue	chemicals in food
smell	nose	chemicals in air
touch	skin	receptors in the skin

Nerves

The body has a complicated network of branching **nerves** (like a network of wires) that carry electrical signals (impulses). These signals start in a sense organ, for example the skin and travel along nerves to the **central nervous system** (the **spinal cord** and the **brain**). A signal is then sent back along different nerves to the muscles, for example the arm muscles, to cause a **response** to the original stimulus.

3. How is the brain protected?

4. If a person breaks their neck in a fall why might they be paralysed?

FIGURE 1: The human nervous system. What does the central nervous system consist of?

brain
central nervous system
spinal cord
nerves

Did You Know...?

Here are some amazing nerve facts.

1. How many nerve cells are there in the body? Hundreds of billions (including 100 billion in the brain alone).

2. How fast do nerves work?
The fastest signals, for example those involved in sensing pain, travel at more than 100 m/s.

3. How is the brain 'wired' into the body? By the all-important spinal cord. It has 31 pairs of nerves branching out along its length. These nerves reach out to all parts of the body.

How Science Works

Sensory deprivation is when the sense organs are minimised in their effect. This allows a deep level of relaxation. To achieve this a person is floated in a liquid in a dark, soundproof container. **HSW**

Nerve pathways

The flow diagram shows the route (pathway) of an impulse.

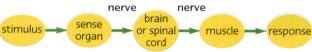

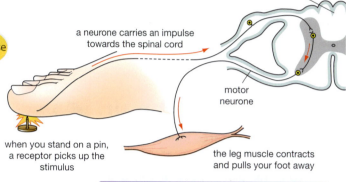

spinal cord

a neurone carries an impulse towards the spinal cord

motor neurone

when you stand on a pin, a receptor picks up the stimulus

the leg muscle contracts and pulls your foot away

5 'Fire, fire! The smell of smoke was detected by James Pond's nose as the boat burst into flames. His brain thought of the options. The muscles in his legs started to work as he ran to the side and dived into the sea.' Identify the stimulus, sense organ and response in the James Pond passage above.

6 The school bell goes off, what will be the students':
a stimulus
b sense organ
c response?

FIGURE 2: The nerve pathway for a reflex action. It is automatic to remove your foot when you step on a drawing pin. The brain is not involved in this action, although pain is detected by the brain. The pathway goes from the stimulus to the spinal column and straight to the muscles. Can you think of an advantage of not including the brain in the pathway in this situation?

Using our senses

7 The following experiments were carried out:
a A pupil was blindfolded. A tapping sound was made by other pupils in different places around the classroom. The pupil had to point to where each sound came from. The pupil was always successful.
b Three bowls of water were set up: hot (45–50°C), warm (25°C) and cold (10°C). A pupil put their left hand in the hot water and their right hand in the cold water. They then put their left hand in the warm water and it felt cold; they then put their right hand in the warm water and it felt hot.
c Ten very small pieces of Plasticine were placed on a table. A pupil squatted down so that their eyes were level with the top of the table. They then used one hand to bring a pin they were holding directly down onto each piece of Plasticine. Using both eyes, they were successful nine times out of ten; but with only one eye open they were successful only three times. (Which eye is best for you?)

For each experiment:
i Explain the results in detail.
ii Indicate the benefit to the species of these actions.

8 Rabbits have eyes at the side of their head, but foxes and owls have them at the front. Explain why, using simple diagrams.

9 Design an experiment to find the smallest difference in temperature that the skin can detect.

Being in control

BIG IDEAS

You are learning to:
- Describe the roles of the brain
- Explain how the brain functions
- Explain how the brain was mapped

The amazing brain

Although all vertebrates (animals that have backbones) have **brains**, the human brain is special. It gives us the power to think, plan, speak and imagine. It is truly amazing.

The brain is a delicate structure that is protected by the bony skull and fluid that forms a layer on the inside of the skull. A boxer can become 'punch drunk' after a punch to the head. This is because their brain has knocked against their skull.

1 Where is the brain in the body?

2 How is the brain protected?

FIGURE 1: What is meant when a boxer is reported to be 'punch drunk'?

Did You Know...?

The brain never actually 'sleeps'. While you are fast asleep at night, your brain is busy.

The role of the brain

The brain carries out a huge range of tasks – many at a fantastic speed. It controls body temperature, blood pressure, heart and breathing rate. It accepts **inputs** from the environment and considers what to do. It gives **outputs** to the muscles to cause movement. It lets you think, dream, reason, remember and experience emotions. All these things are carried out by a structure that weighs only about 1.4 kg – a bit less than the weight of a bag of flour!

3 What cells carry information to the brain?

4 Which tissue does the brain send information to in order to cause movement?

Did You Know...?

The brain is protected by three membranes called the **meninges** that lie just beneath the skull bone. The membranes contain a watery fluid and it is this fluid that cushions the brain from knocks and jolts.
The meninges can become inflamed and this is where the name of the illness meningitis comes from.

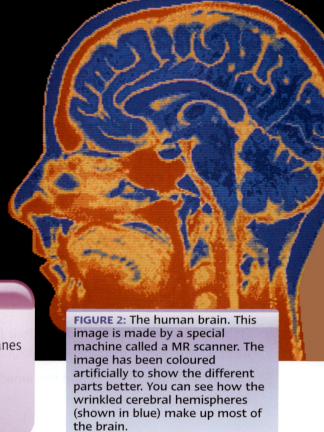

FIGURE 2: The human brain. This image is made by a special machine called a MR scanner. The image has been coloured artificially to show the different parts better. You can see how the wrinkled cerebral hemispheres (shown in blue) make up most of the brain.

Parts of the brain

In order to understand behaviour we need to know about the brain's structure and function. Study Figure 3 to answer the questions.

5 Why does a person 'see double' when they have drunk too much alcohol?

6 A person sees a red car. Why might they think it is a post van?

Mapping the brain

Scientists have **mapped** the brain using evidence from **stroke** victims and using **electrodes**.

- In a stroke part of the brain is damaged because the blood vessels that supply it are blocked or leak blood. In many stroke cases damage is found in the right part of the cerebral hemispheres and this causes a decrease in movement on the stroke patient's left side.

- Electrodes can also be placed inside the brain of animals and impulses sent to see which part of the body moves. This does not hurt the animal.

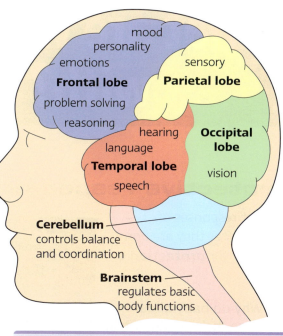

FIGURE 3: The main parts of the brain and some of their functions. What part of the brain is affected by alcohol?

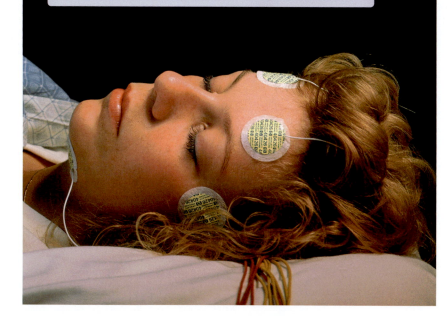

FIGURE 4: Using electrodes to measure brain waves. The sensor pads on the electrodes placed on the patient's scalp pick up the very faint electrical signals that are always passing around the brain. These signals are displayed on a screen as a series of wavy lines. The shapes of the waves change depending on whether the brain is fully alert, thinking, drowsy or in a deep sleep.

How Science Works

Scientists were helped in mapping the brain by studying a patient called Phineas Gage. He had been involved in an accident in which a metal bar went through his brain and skull. His changes in behaviour gave an insight to what the damaged parts controlled.

7 A stroke victim cannot move their left hand. Explain how this fact can be used to map the brain.

8 Electrodes are put into rats' brains and used to map the brain. Give the case for and against the use of rats to map the human brain.

9 Compare the intelligence of the brain to a computer.

What are we born with?

BIG IDEAS

You are learning to:
- Describe the reflexes we are born with
- Explain what a reflex is
- Compare innate and learnt responses

Protective responses

The first responses of animals after they are born are either for **protection** or **food**.

1. Why do baby birds open their mouths when their mother arrives in the nest?

2. Why do rabbits and young birds run for shelter when a shadow goes overhead?

FIGURE 1: A baby bird opens its mouth when its mother returns with food. A young bird that has just left its nest runs for shelter when a shadow goes overhead. Baby rabbits run to their burrow if they are disturbed.

Did You Know...?

The reason a baby wets its nappy is that when its bladder is full a message is sent to release the contents.

As a baby gets older it learns to override this reflex – and becomes 'potty-trained'.

Reflexes

We do not need to learn how to carry out **reflexes** – we are born with them (they are **innate**). Examples of reflex actions are:

- blinking
- sneezing
- hiccuping
- being sick (this removes unwanted material from the stomach).

Anita eats a sandwich and it goes down the 'wrong way'. She quickly starts to choke which brings the food back up. This is called a reflex action. It is a very rapid response to a **stimulus**.

FIGURE 2: Sneezing is a reflex action. Do you plan to sneeze or does it just happen? Are we born with reflexes or do we learn them?

... food ... innate ... learnt response

3 What reflex actions are carried out in the following cases:
  a dust getting into a person's eye
  b pepper getting up a person's nose
  c a person's hand touching a very hot dish.

4 Why must a person get food out of their windpipe quickly?

What responses are animals born with?

Reflex actions are one of the types of responses we are born with. Without them an animal could not survive, as it would be as vulnerable as when it was born. It will learn new behaviour as it grows up.

Some reflexes help the body function.

A dog produces saliva when it smells food (reflex action). This is not thought about. It may later jump up or bark when you go to a certain cupboard at a certain time, as it expects you are going to give it food (**learnt response**).

The main differences between reflex actions and learnt responses are shown in the table below.

Reflex action	Learnt response
very rapid	slower
protective	not all responses are protective
little affected by learning	improved by repetition
not thought about	processed by the brain

How Science Works

Case study: Shadows
A baby chick will run for shelter when a shadow moves overhead. After a while they stop running away from shadows of clouds, but still run away from bird shadows.

a Suggest how they have learnt not to respond to the shadow of clouds.

b Why do they run away from shadows when they are born?

c What benefit to the species is this change in behaviour?

Case Study: Bird song
Baby birds are born with the ability to make sound. They can make the notes that are typical of their species but they do not have the phrasing. As they are reared, they learn the phrasing and so are able to produce the normal song of the species. If they are reared for the first six months without other birds, however, they may never learn the phrasing.

a Use the information above to suggest how the young birds have learnt the phrasing of the song and why the ones reared alone never learn it.

b Suggest why a song is needed and what problems may be caused if the song is not phrased correctly.

Learning behaviour in animals

Watching the parents

All offspring need to **learn** from observing their parents.

FIGURE 2: Ducklings follow their mother from a very early age. Can you suggest why?

FIGURE 1: A toddler learning from its parent. Why do you think babies need to copy their parents?

Trial and error

A young lion cub learns when it moves away from its mother. It might try to eat a wasp and get stung. It will not attack wasps again. It has learnt by trial and error. The cub learns how to creep up on small animals and catch them. It learns how to hide and stalk its prey. By **play-fighting** with its siblings it learns vital skills it will need if it gets into a real fight when it is older.

FIGURE 3: Apart from sheer enjoyment, what is the more serious reason that these young lion cubs are play-fighting?

1. Why does a lion cub leave wasps alone after it has been stung once?
2. Why is it important for a lion cub to learn to hunt at a young age?

Dog obedience training

When an untrained dog is given the command 'sit' it does not understand what is meant. To **train** the dog to sit the trainer gives the voice command 'sit' and pushes the dog's back gently down until the dog is in the sitting position. The dog will begin to associate the voice command with the sitting action. The trainer is **modelling** the

Did You Know...?

Imprinting is a type of learning that happens in the first few hours of life in birds and mammals. The young animal forms a bond with the first large, moving object that it sees after it is born.

An Austrian scientist called Konrad Lorenz carried out an imprinting experiment using greylag geese. He was the first thing that they saw after they had hatched – they followed him everywhere he went, as if he were the mother goose!

Did You Know...?

Can you think why you sometimes see crows sitting on scarecrows in farmers' fields?

After a scarecrow has been in a field for a while the birds learn that it does not harm them.

... behaviour ... learn ... model ... motivate

correct behaviour for the dog. The dog might remember next time it is asked to sit, but there is no particular reason for it to **obey**.

If, however, the dog is given a biscuit every time it sits, the trainer is **motivating** it to sit when it is asked. This type of training is called **positive reinforcement**. If the reward is stopped after a while the **behaviour** associated with it, i.e. sitting, will also stop. A variable period between rewards prevents the behaviour stopping. The animal thinks it will be rewarded again soon so carries on with the behaviour.

This is the principle used to make people play on a slot machine. They keep thinking they may win.

FIGURE 4: 'Sit!' What type of training is often used to teach a dog to obey commands?

3 Why do trainers not give a food reward to an animal every time it does a trick?

4 Suggest how you could train a dog to jump through a hoop.

'Negative' reinforcement

Nowadays, animals are rarely trained by punishment alone. The idea of whipping animals so they cower, using the threat of pain to make them jump hurdles is thought to be cruel and build up stress in the animal. Poor behaviour is still punished:

- A car driver caught speeding gets a fine.
- A dog jumping up is pushed down with a strong 'No'.
- A dog pulls so the owner changes direction quickly to take control of the dog.

It is important to seek advice to train dogs and to be consistent in your actions. Some dogs appear to misbehave because they are 'bored' and by giving them lots of exercise and stimulation, they stop the poor behaviour.

5 Why is it important to ask a trainer when starting to train a dog?

6 Explain why bored animals may misbehave.

Is this really training?

These cases of extreme cruelty to animals are very rare now. Most animal trainers realise that it is much easier to train a happy and relaxed animal that trusts them; rather than bully an animal into doing something it either does not understand or is not able to do.

7 Suggest why the brown bear in the case study above beat its head against the bars of its cage.

8 Do you think positive reinforcement or negative reinforcement training methods should be used on animals? Give reasons for your choice.

Case Study – The use of negative reinforcement training in circuses

In the past, a few circuses have been accused of using brutal training methods on their animals. Whips were used to make big cats cower, screws were hidden in walking sticks and electric probes were also used to prod animals. These are all examples of negative reinforcement methods. An ex-circus employee related how a little brown bear had trouble balancing on a high wire. She was beaten with metal rods until she screamed. She became so neurotic that she beat her head against her small cage. She finally died. The bear was placed in a very stressful situation. It was not able to balance on the wire probably because it was so terrified of what its trainer would do to it and every time it failed to balance it was beaten.

Aggression

BIG IDEAS

You are learning to:
- Describe the signs of aggression
- Explain the causes of aggression
- Link human aggression to animal aggression

'I just blow up'

When a person loses their temper they may start to shout, throw things, lose control or even hit someone. This can be very concerning to anyone nearby. A male red robin flies at anything that is coloured red that enters its **territory**.

Animals show **aggression** in many different ways. They rarely want to fight unless they have to as they might get injured.

1 How does a cat warn off a dog?

2 Why does a robin attack a red puppet hung from a washing line in a garden?

FIGURE 1: Animals show aggression in different ways such as roaring, beating their chests, increasing their size by rearing up, charging and pushing.

... aggression ... group aggression ... physical ... psychological

How do humans show aggression?

- **Physical** violence – for example slapping, hitting, kicking, stabbing, biting.
- **Psychological** threats – for example name calling, malicious teasing, spreading rumours, encouraging rejection by others.

FIGURE 3: Aggressive behaviour takes many forms in humans.

Alcohol seems to increase the chances of **group aggression** as a person becomes less able to control their **responses**.

3 Give **two** reasons why one person might attack another.

4 Suggest **two** ways in which a person can control their aggression.

What causes aggression?

Studies into the causes of animal aggression have helped us to understand why some humans show aggressive behaviour.

At a basic level, aggression in humans may be physical or psychological, whereas animals may threaten or actually attack. The cause of aggression may be one or more of the following.

- Defence of territory – this may be the garden or the home.
- Establishing dominance – an animal defends its area and mates or shows aggression to take over a new area.
- Parent(s) defending young.
- Self-defence – an animal is cornered and reacts aggressively to fight for its life.
- Re-directed aggression – an animal shows aggression to animals near its area even though they are no threat.
- Learned aggression – when an animal is successful in getting what it wants and learns to continue to behave in an aggressive way.
- Attack by predators.

Animal aggression has parallels to human aggression. The causes of aggression in animals and humans depend on the circumstances, the **trigger** (**stimulus**) and how useful the behaviour is to the animal.

5 Why is it not advisable to approach a female deer when she has a fawn (young deer)?

6 Why must you always leave an escape route for a wild animal if you are trying to observe it?

7 Give **two** similarities between human and animal aggression.

8 Use your understanding of the circulatory and breathing system to suggest why a hormone called adrenalin is released when you become aggressive. Adrenalin increases breathing and heart rate.

9 Some people breed fighting dogs, for example bull terriers. Explain how these animals may have become so violent. Why would these animals not be accepted in the wild?

How Science Works

Animals do not usually kill members of their own species; most live in groups and get on well together. Aggression is shown to obtain resources, for example a mate, territory and food. But it has a cost. The animal uses up energy, can be killed or injured, and wastes time fighting that could be used for mating or finding food.

Animals will generally avoid fighting larger and more powerful members of the group. A pecking order may exist, where the animals know their place and those lower down the pecking order will cower when threatened by those a lot higher. This reduces the likelihood of animals lower down being hurt, as they no longer appear to pose a threat to those higher up.

Explain what a pecking order means. Why is it beneficial to a species?

HSW

How do I learn?

BIG IDEAS

You are learning to:
- Interpret results
- Explain why we need memory

How do you learn?

Imagine you have a test tomorrow and you want to **learn** for it. You sit down and read through your notes and then ask somebody to test you. The **motivation** is you want to do well. The learning may only be **short term** so you might not remember the facts next week.

1. Which organ in the body stores memories?

2. Which type of memory, short term or long term, do you improve when you read your notes?

FIGURE 1: What is learning?

The brain and learning

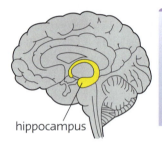

hippocampus

FIGURE 2: There is no single 'memory centre' in the brain, instead many parts work together to store memories. A curved part of the brain called the hippocampus is thought to be important in changing short-term memories into long-term memories that can be recalled weeks or months later.

How Science Works

Parkinson's disease is a disease of the central nervous system that gets progressively worse. Scientists have found that it is related to the break down of nerve centres deep within the brain. Their research has shown that the level of chemicals that carry the message between nerves in the brain is reduced.

You are unable to learn unless you have a **memory**. It is recognised that memory occurs when the **brain** is **stimulated**. Stimulation of the brain causes changes in its nerve cell (**neurone**) pathways. The more times a person repeats the process over a period of time the more a new nerve-cell **pathway** for the message forms in the brain. After a while you do things without thinking as the pathway has been established. Scientists think these pathways can explain short and **long-term** learning patterns.

Did You Know...?

Severe or frightening accidents can cause memory loss of events that took place minutes or hours before the accident.
Emotional stress may result in a person wandering far from home without any knowledge about themselves.

- Short-term learning – when the pathways begin to be established.
- Long-term learning – when the changes in the pathways become permanent.

3. Why can you only learn if you have a memory?

4. Suggest why stroke victims may lose their short-term memory.

FIGURE 3: There are many methods of learning. But they all work by turning short-term memories into long-term memories – hopefully!

... brain ... learn ... long term ... memory ... motivate

Measuring how quickly I learn

A person may learn using their short-term memory and then quickly forget what they have learnt. During short-term memory the brain does not form new nerve pathways straight away – it takes continual **reinforcement** over a long period of time to embed the memory. When a person loses their memory it is the short-term memory that takes the longest to return. In old people some types of stroke result in them not remembering what has happened recently.

Learning is difficult to measure. You are going to investigate how quickly you learn a sequence of letters and numbers.

Your teacher will provide you with the apparatus that you may need for your investigation.

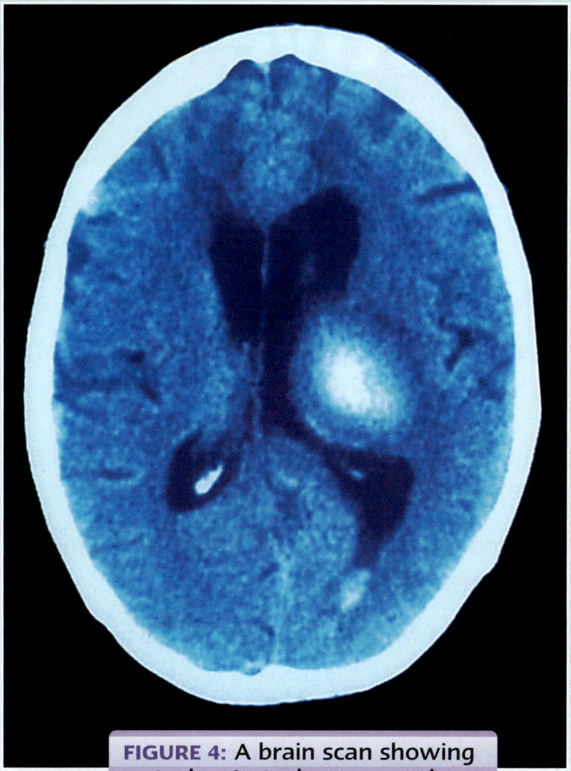

FIGURE 4: A brain scan showing a stroke. A stroke occurs when the blood supply to part of the brain is interrupted, leading to the death of some brain tissue. The white area is the stroke.

Method:

1 Take one sheet of plain paper and a stopwatch.

2 Write the letters of the alphabet in reverse order starting from R and finishing with H. Time how long it takes you to do this.

3 For every mistake you make add 3 seconds.

4 Repeat **steps 2** and **3** 12 times. Each time fold the paper so your previous attempt is covered.

5 Now repeat **steps 1** to **4** this time writing the numbers 18 to 8 in reverse order.

Questions

1 Record your results in a table and then using squared paper plot a graph of your results.

2 What pattern is shown by your results?

3 What do your results show about learning?

4 Is there any difference between your results and those of the rest of the class?

5 Suggest reasons for differences in learning.

6 Design an experiment to see if Year 9 students learn better by looking at eight pictures of objects or eight written names.

7 Is the speed of your reactions innate, learnt or a mixture of both. Indicate what evidence you would need to collect to find this out and how you would use it.

Effective learning

BIG IDEAS

You are learning to:
- Describe some different methods of learning
- Explain how to learn effectively
- Explain what mind mapping is

The driving test

In the case study, the type of learning that Mohammed is using is called **passive** learning.

1 Why does Mohammed read through the code several times instead of just once?

2 Suggest why Mohammed might lose his concentration.

3 How could Mohammed make his learning more effective?

> **Case study – Learning for my driving test**
>
> Mohammed says "It is my driving test tomorrow. For part of the test I am tested on the *Highway Code*. Tonight I am going to sit down and learn the code by reading through it several times. It is hard work just reading as I keep losing my concentration and thinking of something else. How can I make my learning more effective?"

Learning methods

- When we are young we learn by **imitating** (copying) our parents. This is called **visual** learning.
- Learning also occurs by doing and experimenting, for example building sand castles and riding a bicycle. This is called **kinaesthetic** (or **active**) learning.
- Finally we learn by reading and listening – this is called **auditory** learning. Most people have a listening attention time of about 20 minutes, after that their minds start to wander.

4 a A science teacher carries out a practical. Which type of learning is taking place in each step, **A** to **C**, of the investigation?

    A The students read the practical method and then listen to their teacher who is giving instructions.

    B The teacher demonstrates the practical.

    C The students do the practical.

  b Why are stages **A** and **B** needed?

How Science Works

A photographic memory is the ability to recall images, sounds or objects with great accuracy and abundance. Some people suffering from **autism** have extraordinarily good photographic memories. **HSW**

Embedding learning

> **Case study – Learning the part of 'Captain Jack'**
>
> An actor with the part of Captain Jack goes on stage having learnt his lines. He finds it easier to learn his lines beforehand by practising them with a partner rather than just reading them through to himself. This makes his learning active.
>
> - If the actor performs the play only a few times he will forget his lines after a few weeks. His learning will not have been **embedded**.
>
> - If he performs his role on stage every day for more than six weeks his learning will be embedded and he will remember his lines for months or even years to come.

... active ... auditory ... autism ... embed ...

How do you learn?

Do you learn better by seeing or reading? You will carry out a simple investigation.

You will be given a pack of 12 cards with the object to remember what is written on them. You will have 5 minutes to learn these words.

You will then be asked to write down as many as you can remember.

You will now be given 12 pictures of objects and will be given 5 minutes to remember them.

You will be then asked to write as many as you can remember.

5 **a** Which method is the most successful for you?
b Evaluate the method and results of this investigation.
c How valid is the conclusion you have made about your learning style?

Mind mapping

Your experiments can be used to help you revise.

A **mind map** is an important technique to improve a person's ability to make notes in such a way to make the information more easily remembered. It does this by linking information together in a sequential way and also by using illustrations so that the brain has a visual stimulus as well.

Mind maps help people to solve problems using a creative approach. They are also very valuable in revising for exams.

However mind maps do not suit everyone. This is because everyone learns in a different way. For example, some people find it easier to use a **spider diagram** to organise their notes.

Figure 1 summarises the main methods of learning.

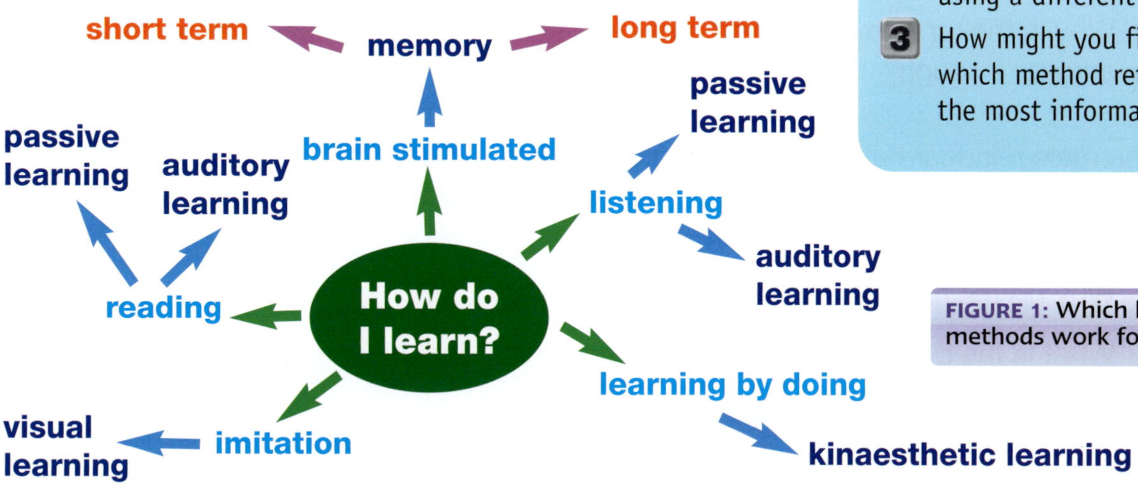

How Science Works

Learning a poem
You have 10 minutes to learn a poem of eight lines. Split the group.

- At least 12 will learn it on their own.
- At least 12 will work in pairs reading out the poem and being corrected/helped.

The people will perform the poem and be judged on the number of mistakes made.

1 What conclusions can be made from the results? Is the sample large enough?

2 Why would it be appropriate to repeat the experiment but swap the tasks around using a different poem?

3 How might you find out which method retains the most information?

HSW

short term ← memory → long term

passive learning
auditory learning
brain stimulated
reading
How do I learn?
passive learning
listening
auditory learning
learning by doing
kinaesthetic learning
visual learning
imitation

FIGURE 1: Which learning methods work for you?

1 A cigarette produces toxic chemicals when smoked. Match the chemical to its effect.

Chemical	Effect
Tar	Addictive drug
Carbon monoxide	Can cause cancer
Nicotine	Reduces the oxygen carried in the blood

2 Theo heats water in a glass beaker using a Bunsen burner and a tripod. When he finishes his investigation he clears away. He picks up the tripod but it is too hot. He drops it straight away.

a What is the name of this type of action?

b How did the message get from Theo's hand to his central nervous system?

c What is the advantage of the reaction being very quick?

3 A survey was carried out to find the effects of smoking on health. The results are shown below.

Number of cigarettes	Number of deaths/ 100 000 per day from lung cancer	Number of deaths/ 100 000 people/year from bronchitis-related diseases
0	3	2
1–14	6	20
15–24	50	35
25 and above	200	59

a Which part of the lungs are affected by bronchitis?

b What relationship is there between the number of cigarettes smoked and the number of deaths from **i** lung cancer **ii** bronchitis-related diseases?

c How do the results for the effect of smoking differ for lung cancer and bronchitis-related disease?

d Why is the tar from a cigarette not swept out of the lungs if a person has a smoker's cough?

4 The male reindeer has his own group of females to breed with. Occasionally he is challenged for the herd by another stag. They roar at each other, charge, push each other, but do not fight to the death.

a Why do the reindeer not fight to the death?

b What is the benefit to the species of one male deer having many females?

c How do the young males learn to fight?

5 Diagrams of nerve cells are shown below.

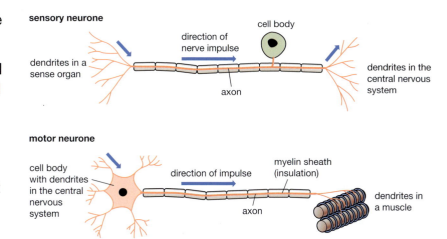

sensory neurone

direction of nerve impulse

cell body

dendrites in a sense organ

axon

dendrites in the central nervous system

motor neurone

cell body with dendrites in the central nervous system

direction of impulse

myelin sheath (insulation)

axon

dendrites in a muscle

a Explain why a nerve cell is sometimes described as being like a wire.

b What is the role of a nerve cell?

c What is meant by 'input nerve cell' and 'output nerve cell'?

6 Toni decided to give up smoking having smoked for several years. The following results are seen over a five-year period.

Time since giving up smoking	Effect on the body
1–2 hrs	The oxygen level in the blood increases to a normal level. The risk of a heart attack starts to fall.
24 hrs	Carbon monoxide leaves the body. The lungs start to clean out mucus.
48 hrs	Breathing becomes easier. Energy levels increase.
2–12 weeks	Circulation improves. Walking and exercise become easier.
3–9 months	Breathing problems, coughing and shortness of breath improve. Lung efficiency increases by 5–10%
5 years	Risk of lung cancer falls to half that of a smoker. Risk of heart disease returns to normal.

a Explain why she was less breathless and her energy levels increased after a few days.

b Explain why coughing and shortness of breath have improved after 6-9 months.

c Explain in detail why a person becomes irritable when they stop smoking.

7 Pavlov observed that dogs salivated when they saw or smelt food. He rang a bell before they received the food and found after several trials that they would salivate when he rang the bell.

a What benefit is it to the dogs to salivate when they smell food?

b Use the information to explain the difference between a reflex action and a learnt response.

c Use the above idea to design an experiment to see whether dogs can tell the difference between red, green and blue light.

Learning Checklist

4

☆ I know drugs affect my behaviour. — page 8

☆ I know the chemicals found in a cigarette. — page 12

☆ I know eyes detect light, ears detect sound and the nose and tongue detect chemicals. — page 22

☆ I know the brain is in the head and is protected by the skull. — page 24

5

☆ I can explain what a depressant does. — page 8

☆ I can explain what a stimulant does. — page 8

☆ I know alcohol is a depressant. — page 10

☆ I know the harmful effects of smoking. — pages 12–13

☆ I know the role of the brain. — page 24

☆ I know why a reflex action is important. — page 27

☆ I know that learning is improved by repetition. — page 32

6

☆ I know what tolerance and addiction mean. — page 9

☆ I know the symptoms of withdrawal. — page 9

☆ I can explain why smoking causes heart disease. — page 13

☆ I can give the pathway for a reflex action. — page 23

☆ I can explain how learning by reward and learning by punishment occur. — pages 28–29

7

☆ I can give a reasoned argument on whether alcohol should be banned. — page 11

☆ I can give a reasoned argument on the effects of cannabis. — page 17

☆ I can give a reasoned argument on which the worst drug is. — page 19

☆ I can explain how a nerve cell is specialised for its role. — page 23

☆ I can design experiments to investigate how complex learning occurs. — page 33

☆ I can explain how learning changes the behaviour of an animal. — pages 28–29

8

☆ I can give a detailed explanation of how smoking causes breathlessness. — pages 12–13

☆ I can identify the evidence that needs to be collected to show the effect of drugs and explain how it is used. — page 19

☆ I can explain the benefits and cost of aggression to a species. — page 31

Topic Quiz

1. What drug is found in coffee?
2. Name **three** dangerous chemicals found in a cigarette.
3. Name **three** organs affected by alcohol.
4. Why is it difficult for a smoker to remove mucus from the lungs?
5. What is meant by withdrawal symptoms?
6. Why can smoking increase the risk of amputations?
7. What are our five senses?
8. Which senses are well developed in rabbits?
9. Give **two** roles of the brain.
10. What is the difference between a learned response and an inherited response?

True or False?

If a statement is false then rewrite it so it is correct.

1. Tar causes addiction.
2. Alcohol is a depressant.
3. Drinking half a pint of beer will not affect your reactions.
4. Cannabis does not cause long-term diseases.
5. A depressant makes you feel upset.
6. The ears detect vibrations in the air.
7. Riding a bicycle is a reflex action.
8. Identical twins have the same memories.
9. A female animal shows aggression if she is approached when she has offspring.
10. Nerve cells are surrounded by a fatty layer.

Numeracy Activity

The table shows the average brain weights (grams) for different animals.

Use the table to answer the following questions.

1. Which animal has the largest brain?
2. Which animal has the smallest brain?
3. Suggest why a whale has a bigger brain than a human.
4. What evidence is there that mammals are more intelligent than reptiles?
5. What evidence is there that large mammals are more intelligent than small mammals?
6. Give one reason why the statement in Q5 may be wrong.

Species	Weight (g)
adult human	1300–1400
elephant	4783
polar bear	498
lion	240
sheep	140
sperm whale	7800
gorilla	465–540
rabbit	10–13
rat	2
turtle	0.3–0.7
viper	0.1
alligator	8.4
shark (great white)	34
bull frog	0.24

Revision Activity

Draw a mind map to remember this topic. Remember to include illustrations for each main point.

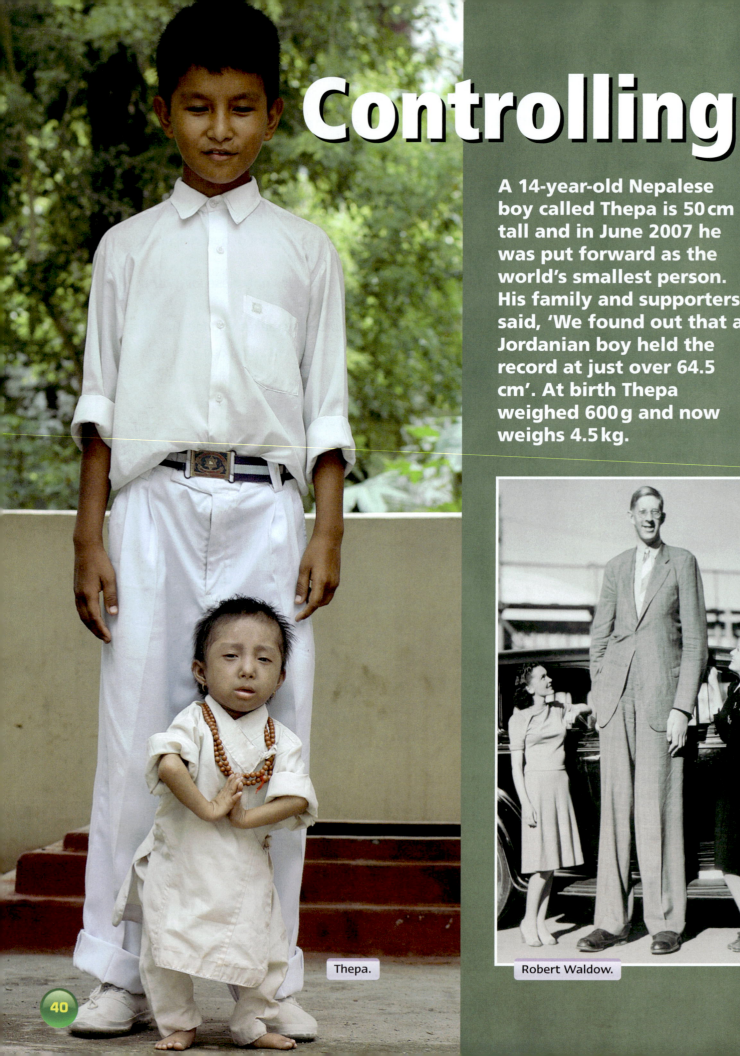

Controlling

A 14-year-old Nepalese boy called Thepa is 50 cm tall and in June 2007 he was put forward as the world's smallest person. His family and supporters said, 'We found out that a Jordanian boy held the record at just over 64.5 cm'. At birth Thepa weighed 600 g and now weighs 4.5 kg.

Thepa.

Robert Waldow.

growth

BIG IDEAS

By the end of this unit you will be able to describe variation and explain why it is important. You will be able to explain how selective breeding works and why it is useful. You will have used scientific ideas to explain things that happen.

- The tallest person ever was Robert Waldow who grew to 2.7 m (8 ft 11 "). He died from an infected foot.

- The tallest living man is Xi Shun at 2.3 m (7 ft 8 ").

- In 1976 Sandy Allen was declared the world's tallest living woman at 2.28 m (7 ft 7.25 "). Sandy underwent surgery at the age of 22 to stop her growth. She can no longer stand upright and is confined to a wheelchair because her body weight has caused her back to arch.

- The heaviest adult is Manuel Uribe who weighed 270 kg before he went on a diet to save his life.

- The person with the longest hair is Xie Qiuping whose hair was 5.55 m long (18 ft 5.54 ").

There are many facts about the differences between individuals. *The Guinness Book of Records* is a good place to look.

What causes differences in the heights of people?

A person inherits information from their parents and this determines their height. The environment also has an affect, for example a healthy diet helps a person to grow to their full height.

In the case of circus dwarfs, they are unable to produce a chemical called human-growth-hormone because the gene that controls its production is not working properly. It is this hormone that controls growth in a person.

- Human-growth-hormone has been used on children who have stunted growth because their bodies are unable to make the hormone.

- Human-growth-hormone is produced throughout a person's lifetime, but at higher levels during adolescence. It stimulates growth in children and increases their metabolic rates.

In order to be able to treat people the hormone was originally extracted from dead people but now it is produced using genetic engineering.

- In 1959 stunted children were given the hormone. Boys without the treatment grew to a height of 1.3 m by the age of 18, whereas those treated with the hormone reached a height of just over 1.8 m. Recently it has been used to reverse muscular wasting in people suffering from AIDS.

- Studies have shown that a raised level of human-growth-hormone causes swelling of the soft tissue in the body, abnormal growth of the hands, feet and face and high blood pressure.

What do you know?

1. What is the difference in heights of the smallest living person and the tallest living person in the world?

2. How does the amount and quality of a person's food affect their height?

3. What health problems could Manuel suffer from by being overweight?

4. Why did Sandy's back start to arch?

5. What is 'AIDS'?

6. What does 'metabolic rate' mean?

7. Why do levels of human-growth-hormone rise during adolescence?

8. Why were trials carried out on human-growth-hormone before it was used in humans?

9. What problems might be caused if human-growth-hormone treatment was addictive?

10. What evidence is there that height is controlled by genes?

11. What problems do you think a very tall person of height 2.25 m (7 ft 7 ") would have – socially and biologically?

12. Sandy has spent part of her life touring the country to help other tall people accept themselves. Why might other people find very tall people unacceptable?

Why are we different?

BIG IDEAS

You are learning to:
- Recognise differences caused by environmental factors
- Recognise differences caused by inherited factors
- Explain how differences in people are caused by environmental factors and genetic factors

Why are we different?

One cause of differences in people is their **environment**. For example, diet, health and amount of exercise. The other major cause of differences in people is **inheritance**. This causes differences in people as a result of **genetic factors**. We can use the word **variation** to describe differences in people.

- Parents pass on information in the **nucleus** of their **sex cells** to their offspring.
- The offspring inherits one set of information from the mother and one set from the father.
- Sometimes there are clear **traits** that run in families such as shape of the nose, double chin or freckles.

1. Name **one** characteristic that you have inherited from your parents and one caused by the environment.

2. Look at the photograph of the sow (mother pig) and her piglets below and indicate one feature that might be inherited.

FIGURE 1: What trait runs in this family?

What is in a nucleus?

In every nucleus there are long thread-like structures called **chromosomes**. Each chromosome is made up of many units called **genes**. Genes determine the characteristics of an individual, for example hair colour. A gene is made of a chemical called **DNA** and this contains the 'blueprint' (recipe) to make an individual.

... chromosome ... DNA ... environment ... environmental factor ... gene

3 Explain what the difference is between a gene and a chromosome

4 Produce a simple diagram to show the link between a cell, nucleus, chromosome and gene.

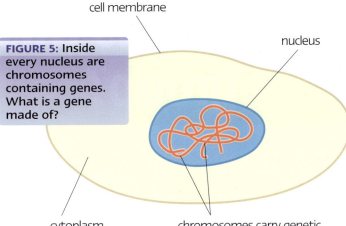

cell membrane

nucleus

FIGURE 5: Inside every nucleus are chromosomes containing genes. What is a gene made of?

cytoplasm

chromosomes carry genetic information in genes

The importance of being different

Animals and plants of the same species have many similarities as they have genes that are common to the species, but they still show differences as they have different forms of the gene.

Some differences benefit the species, for example:

- Not all rabbits are killed by myxomatosis. Some rabbits are resistant and will survive an outbreak.
- Not all mosquitoes are killed when a pesticide called DDT is used.
- Peacocks with the best display of feathers are most likely to attract a mate.

Some differences are not beneficial, for example:

- Albino (pure white) giraffes do not survive long in the wild.
- Antelopes that run slower than the herd do not survive.

5 Explain why albino giraffes do not survive.

6 Explain why variation is important.

7 Explain why when DDT is used the numbers of mosquitoes first go down but then start increasing again.

Did You Know...?

Every cell in your body (except the sex cells) contains all the information needed to make a new individual. The reason that a muscle cell forms in muscle tissue is that only the genes that make muscle cells are switched on.

Did You Know...?

DNA finger-printing is used to identify criminals.
Any material that contains DNA, for example a single hair, a tiny drop of blood or even skin cells left at a crime scene can be duplicated and then matched to DNA from a suspect.

Case study – Genetic and Environmental factors are present

In the 19th century, studies were carried out on the children from poor families. The families had little food, no books and the children worked for long hours. They compared their ability to the children from rich families, professionals and the clergy. They concluded that the children from the upper classes were more intelligent and that this was inherited from their parents and that the children and parents from poor families lacked intelligence. This was used by some people to claim that it was not worth educating the poor. This has now been discredited.

8 Give another explanation as to why poor children performed badly.

9 Suggest how they could have carried out a study to disprove their ideas.

... genetic factor ... inheritance ... nucleus ... trait ... variation

How tall is this group?

BIG IDEAS

You are learning to:
- Describe about the methods used to measure variation
- Explain the difference between reliability, accuracy and precision

What can I measure?

There are many differences in people. Some of these differences are easily **observable** and so they can be easily **recorded**. Look at the table below that records the number of students in a class grouped by whether they can roll their tongue and by their hair colour.

FIGURE 1: What observable differences are there in the students of this class?

	Dark hair	Fair hair	Tongue roller	Not tongue roller
girls	12	7	14	5
boys	8	6	11	3

1 **a** Which is most common in the class, tongue rolling or not tongue rolling?

b What hair colour is least common in the class?

Case study – How large should a sample be?

Julia's group were calculating the most suitable sample size to work out the hand span of year 9 pupils from a group of 30.

Sample	Number of pupil chosen at random	Average handspan (mm)
1	4	172
2	8	186
3	12	180
4	16	185
5	20	183
6	30	182

2 Explain why the sample is chosen at random.

3 Suggest how you might chose the sample at random.

4 Draw a graph of the sample size against average hand spans. Join up the points.

5 Explain what you think an appropriate sample size is and give reasons.

... *accuracy ... observe ... precision*

Growing mustard

You are going to investigate whether mustard grows faster in the dark or the light.

Method:

1 Take **two** petri dishes and place a piece of very wet filter paper on the bottom.

2 Place 30 mustard seeds in each dish

3 Cover both dishes with a clear plastic beaker.

4 Cover one beaker with a black material.

5 Leave the seeds for ten days.

You will decide:

- how many seedlings to measure
- How you will measure them
- How you will select your sample.

Exam Tip!

- Reliability – how close the results are to each other. The more measurement improves the reliability.

- Accuracy – the closeness of the results to the 'real' answer. Accuracy in measurement is improved by the apparatus used.

- Precision – determined by the equipment used, working with complete accuracy and observing a strict set of rules.

Questions

1 **a** Why are the seeds covered with a beaker at the start?
   b How have you made the test fair?
   c State what decisions you have made and explain why you made this decision.
   d What conclusions can you make from your results? How have you compared the results?

2 Explain how you have attempted to make your results reliable, accurate and precise.

What can twins tell us?

BIG IDEAS

You are learning to:
- Interpret patterns in the data
- Explain why identical twin are genetically the same
- Evaluate evidence on identical twins

Differences between twins

Case study – How are Duncan and Reece different?

Duncan was split from his identical twin Reece 10 years ago. Reece went to live in Australia near the beach and Duncan stayed in London where he worked in an office. They both are six feet tall and have brown hair and eyes.

When they meet Duncan sees that Reece is more muscular, his skin is brown and he speaks with an Australian accent. Reece has a scar over his eyes from a rugby injury.

1 What evidence is there that scars are caused by the environment?

2 What evidence is there that eye colour and hair colour are inherited?

FIGURE 1: Identical twins are the same gender.

Identical twins have the same genes

Identical twins have **inherited** the same **genetic** information from their parents. This is because they came from the same **fertilised** egg. Any differences existing between identical twins will be due to the effect of the **environment**.

- There are only slight differences between identical twins brought up together as their environment is similar.
- There are far greater differences between identical twins brought up apart as their environments are different.

3 Why do identical twins have the same genetic information?

4 What environmental factors caused Duncan in the previous case study to have a paler skin and less muscle?

How Science Works

Studies of sets of identical twins brought up apart cannot easily be carried out.

There are records of twins parted during the Second World War and twins adopted separately from homes. However this does not represent a cross-section of identical twins.

... bias ... data ... environment ... fertilise ... genetic

What can data show?

The **data** in the table below show the **mean difference** in measurement between many sets of identical twins raised together and identical twins raised apart. The mean difference is worked out by averaging the heights or masses or head widths of identical twins raised together, and of those raised apart. These means are then compared to work out the difference.

	Height (cm)	Mass (kg)	Head width (mm)
identical twins raised together	1.7	2.0	2.8
identical twins raised apart	1.8	4.8	2.9

5 Explain what the information in the table shows about the inheritance of:
   a height
   b mass in identical twins.

6 Why is the *mean* difference worked out for many sets of identical twins?

The table shows that identical twins raised together and apart have similar head widths as the mean difference is nearly the same: 2.8 mm compared to 2.9 mm. It therefore seems likely that head width is mainly genetically determined rather than environmental.

Twin studies

The problem of studying identical twins brought up apart is that there are few cases to study. Records are based on twins parted during the second world war or those adopted from children's homes separately.

A classical twin study involes:

- Studying twins raised in the same environment
- Identical twins have the same genes
- Non-identical twins share about 50% of them
- Researchers compare the similarity between sets of identical twins to those of non-identical twins
- They assume exact likeness (far higher than 50%) is due to genetics rather than the environment

7 Explain the benefits of studying identical and non-identical twins rather than just identical twins brought up together and separately.

8 Give a reasoned account of the ethics of the case study.

Case study

Identical twins Paula Bernstein and Elyse Schien met at the age of 35. They had been parted at birth and were not aware of each other's existence. Paula said 'We had 35 years to catch up on. Where do you start?'. Elyse had been brought up in Paris and decided to seek her birth mother and found she had an identical twin. She discovered that she had deliberately been separated at birth and they were subject to a unique study on nurture versus nature.

It came from my parents

BIG IDEAS

You are learning to:
- Explain how information is passed on from the parents
- Explain the differnces between sexual and cell division
- Predict the results of genetic crosses

HSW

What does 'inherited' mean?

Each parent provides information contained in the **genes** in the **chromosomes** of the **sex cells**. Each sex cell contains half the full amount of chromosomes. When an egg is **fertilised** the male sperm (sex cell) joins with the female egg (sex cell) so the baby has a complete set of information from the father and from the mother. The baby has **inherited** this information.

Remember only the sex organs produce sex cells. Other body cells divide to form two identical cells. Each has the full set of chromosomes.

1 Where in the sex cells are chromosomes found?

2 A human embryo has 46 chromosomes. How many chromosomes are in
   a a sperm cell
   b an egg cell?

3 Give the similarities and differences between cell division and sexual reproduction

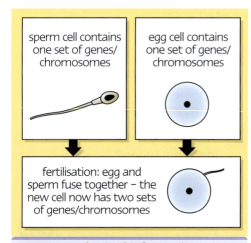

sperm cell contains one set of genes/chromosomes

egg cell contains one set of genes/chromosomes

fertilisation: egg and sperm fuse together – the new cell now has two sets of genes/chromosomes

FIGURE 1: Inherited information. The sex of a human embryo is an inherited feature.

Why has Amanda got freckles?

It is genes that carry the information that determines what sex we are and what we look like.

There are **two** genes for each feature in a cell – one from each parent. If freckles is given the symbol F, then if two freckled parents have children the diagram below might represent what happens.

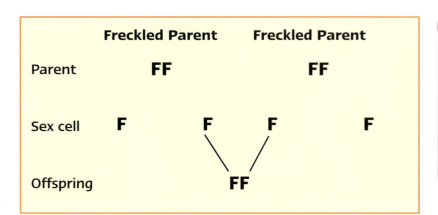

	Freckled Parent	Freckled Parent
Parent	FF	FF
Sex cell	F F	F F
Offspring	FF	

Did You Know...?

Freckles are caused by a brown pigment in the skin called melanin. Sometimes the cells that produce melanin are not evenly spread throughout the skin. So some areas of skin have a darker colour which we call freckles.

... chromosome ... dominant ... fertilised ... gene

The diagram below shows what might happen if one parent has genes for freckles and the other parent has genes for no freckles.

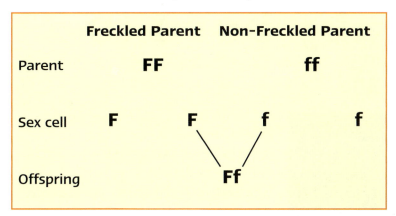

	Freckled Parent	Non-Freckled Parent
Parent	FF	ff
Sex cell	F F	f f
Offspring	Ff	

How Science Works

The DNA of parents can be analysed by scientists to find out if they are carrying any genetic diseases in a recessive form that they may pass on to their children

HSW

All the offspring are Ff and as the freckled gene, F, has the strongest influence (it is **dominant**). They all have freckles.

4 Copy and complete the cross on the right to show what offspring may be produced. The freckled parent is Ff.

Non Freckled sex cell

	f	f
Freckled sex cell F		
f		

5 If the non-freckled parent had been freckled (Ff instead of ff) use the cross diagram from Q4 to show what the offspring would have looked like.

Eye colour

The inheritance of eye colour is more complex. Dominance is sometimes not so clear cut. In some cases, such as eye colour, there is **incomplete dominance**. This means that the gene controlling brown eyes does not exert total dominance, so shades of brown are produced in the offspring such as hazel and light brown. Blue eye colour is **recessive**. If two brown-eyed parents carry the recessive blue gene then they can pass blue eye colour on to some of their offspring.

6 **a** Copy and complete the cross on the right. Let B be the symbol for brown eyes and b the symbol for blue eyes.
   b What colour eyes will the offspring have?

	B	b
B		
b		

7 Sex is determined by the chromosomes the individual has, not the genes. The sex chromosomes are X and Y. A female has the chromosomes XX and a male XY.
   a Complete the following
   b **i** What is the chance of the first child being a boy?
   ii What is the chance of the second child being a boy?
   c Explain why the sperm determines the sex of the child not the egg.

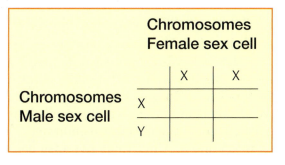

Chromosomes
Female sex cell

Chromosomes Male sex cell		X	X
	X		
	Y		

... incomplete dominance ... inherited ... recessive ... sex cell

Test tube baby born to save ill sister

A Colorado couple has used genetic tests to create a test tube baby that has the exact type of cells desperately needed to save their 6-year-old daughter, Molly.

The case is the first time a couple has screened its embryos before implanting them in the mother's womb for the purpose of saving a life.

Molly has inherited Fanconi anaemia, a disorder that causes the production of bone marrow to fail. Children with this disorder suffer from bleeding and immune system breakdown and usually die by the age of 7 years.

The only effective treatment is to get a batch of healthy bone marrow cells from a perfectly matched sibling to replace the ailing child's faulty bone marrow.

Molly's parents have always wanted to have more children, but have been afraid they too would inherit the disease. A few years ago they learned of a new technique under development called pre-implantation genetic diagnosis. This means Molly's mother's fertilised egg cells can be screened for the presence of the faulty gene.

Lisa Nash, Molly's mother, allowed researchers to test the embryos for compatibility with Molly's cells, but this was not good. Finally, after five attempts, two embryos free of the disease that were also a good match to Molly's cells had been identified. Only one of the two embryos implanted and on Christmas Eve Lisa became pregnant.

After Adam was born doctors saved blood cells from the umbilical cord. They performed the transplant on Molly in September. "Molly was holding Adam in her lap" Lisa Nash said, while the cells dripped through a plastic tube into her chest. "It was an awesome, monumental experience of our life". Afterwards, Molly was playing with her computer in her room and would soon be out of the high-risk period.

Here a life has been saved by scientists. The concern is that the work could point to a future where parents have the option to choose traits in their children, for whatever practical or faddish reasons they have. Do we want a 'design a baby' culture where parents choose the sex of the baby, the hair colour and the eye colour? "It's like buying a new car, where you decide what package you want" said Jeffrey Khan, a Director of the Centre for Bioethics. "It's only because we are still developing the tests that we do not have parents asking for kids who are 6 feet tall."

Assess Yourself

1 How long might Molly have had left to live if she had not had the transplant?

2 Which part of Molly's body was faulty?

3 What does 'fertilised' mean?

4 What does an 'inherited disease' mean?

5 Why did the Nash family want another child?

6 Why was the Nash family concerned about having more children?

7 What is anaemia? What problems does it cause?

8 Explain what a test tube baby is.

9 What test did the scientists carry out on the embryo to check if it could be used to help Molly?

10 Why was there such a rush to have baby Adam?

11 What is bioethics?

12 What problems might be caused by having one child to save another?

13 Give the case for or against parents being able to determine features of their baby.

ICT Activity

Find out from the Internet if there are other examples of babies being conceived to help their brother or sister.

Citizenship Activity

Discuss the problems that could be caused by families if they were able to determine the sex of their children. What problems would be caused if they selected boy babies?

Level Booster

8 Your answers show that you have the ability to communicate scientific ideas and arguments to reflect a range of views.

7 Your answers explain the importance and implications of science in using one child to safe the life of another.

6 Your answers explain the importance of carrying out tests on fertilised eggs before they are implanted.

5 Your answers describe the process of fertilisation using appropriate terminology and describing its implication in producing test tube babies.

4 Your answers describe examples of characteristics that are inherited and of those that are caused by the environment.

Natural clones

BIG IDEAS

You are learning to:
- Describe why some plants have identical offspring
- Explain how cloning can be carried out by scientists
- Discuss the ethics behind cloning

Plants that are exactly the same

You have probably seen a spider plant. It has little plants dangling from the end of '**runners**' coming from the parent plant. All the little plants look the same.

Strawberry plants also spread by runners. The new plants rely on the parent plant until their own root system is fully formed. The roots provide the plant with water.

FIGURE 1: Spider plants and strawberry plants produce identical offspring.

1. How does the spider parent plant support the young plants before they become detached from the parent?

2. Why do strawberry plants spread quickly in an area?

Getting rid of weeds can be difficult

Some plants can be very difficult to remove unless every part of them is dug up.

FIGURE 2: The dandelion can re-grow from small sections of root – each new plant is a replica of the original plant. Ground-elder has underground stems which have to be totally removed – if they are not the weed soon appears again – much to the annoyance of gardeners!

3. Why is it difficult to get rid of weeds such as dandelions and ground-elder?

4. Why do most plants have brightly coloured flowers?

Did You Know...?

Japanese knotweed was introduced into Britain in 1825 from Japan as an ornamental plant. However, it soon became clear that it is an aggressive plant that spreads rapidly and engulfs other plants. It is very difficult to get rid of as it has underground stems that can re-grow. A piece accidentally left in the soil the size of a little finger nail can re-grow.

In the UK it was made illegal to spread Japanese knotweed by the Wildlife and Countryside Act 1981 and it is 'controlled waste' under the Environmental Protection Act 1990. This means it requires disposal at licensed landfill sites.

It usually takes at least three to four years to eradicate the plant using a weedkiller.

Reproduction with no sex

Some plants can produce new individuals without using sex cells. This is called **asexual reproduction**. Each new individual looks the same because it has the same genetic information as its parent plant. The new individual is called a **natural clone**.

Plants therefore have two methods of reproducing.

- **Sexual reproduction** – involves sex cells and produces individuals different from the parents.
- Asexual reproduction – does not involve sex cells and produces individuals identical to the parent.

Look back at the strawberry plants shown in Figure 1. What evidence is there that the strawberry plant uses both methods of reproduction?

5 Which method of reproduction produces variety in the offspring?

6 Suggest which method of reproduction allows offspring to establish more quickly.

Manmade clones

Some plants naturally produce clones. Few animals can produce **natural clones**, and few animals can be cut up and produce new animals. You cannot remake a human from a finger. One way scientists have cloned animals is to take an egg cell and remove its nucleus and replace it with the nucleus from a normal body cell of the animal they wish to clone. This process is called **nuclear transfer**.

7 Greenfly can produce clones quickly. They usually do this in the summer. What is the advantage of reproducing rapidly in the summer?

8 Suggest why there is concern about cloning humans in the same way as Dolly the sheep was cloned.

Sheep A — egg cell taken from sheep A and nucleus removed

Sheep B — cells taken from the udder of sheep B and the nucleus removed

nucleus from sheep B is put into egg of sheep A

egg cell is put into a female sheep to grow

cell grows into a clone of sheep B

FIGURE 3: Cloning an animal. In Dolly the sheep's case a nucleus from her sister's udder cell was used. Would Dolly have looked different if a nucleus from another part of her sister's body was used?

Case study – Dolly the sheep clone

Dolly the sheep was cloned using the method described above. She was identical to her sister sibling who donated the nucleus. In producing Dolly over 270 foetuses had to be aborted. Dolly died aged 6 (sheep of her type are expected to live to about 12 years). It is thought that Dolly's early ageing was caused by using a nucleus from her 6-year-old sibling — so Dolly was genetically already 6 years old when she was born.

Dolly (1996–2003).

9 Science fiction suggests that top scientists could produce cones of themselves. Make the case for and against the use of human cloning. Consider issues concerning the ethics, the rights of the clone and medical benefits.

Dog breeding

BIG IDEAS

You are learning to:
- Explain the meaning of artificial selection
- Explain how artificial selection has been used to breed pedigree animals
- Evaluate the problems of inbreeding

How did the modern dog evolve?

Exam Tip!

Selective breeding or artificial selection means choosing the type of organism to breed and then selecting which offspring to breed.

The domestic dog has been produced by a process called **artificial selection** or **selective breeding**. Man selected the wolf with features that were desirable. (Remember all animals have differences – some of which we like; some we don't.) He bred these animals together. He then selected the best of the offspring and bred these together. Over many thousands of years the wolf was selected to become the domestic dog. Originally the dog was bred as a working or a hunting animal.

1. What features of the wolf are desirable to make a hunting dog?

2. What features did man want to breed out of the wolf?

3. Explain why man bred offspring only with features that were wanted. What did he assume?

Breeding pedigrees

The most common way of breeding **pedigree** dogs is called **line breeding**. This means dogs that are distantly related, for example distant cousins, are bred together. This means both parents carry the desired features. New undesirable features are slowly removed by selecting only dogs with the correct features. In this way the unwanted features are bred out. The puppies are fairly similar. A disadvantage of line breeding is that it can take several generations to get rid of unwanted features.

FIGURE 1: 'Man's best friend is his dog.' How did a wolf become an obedient pet?

How Science Works

Until recently scientists were not sure whether the domestic dog originated from the desert wolf, the woolly wolf or the gray wolf. The use of DNA analysis has provided strong evidence it was the gray wolf.

FIGURE 2: Several generations of pedigree labradors. Why do pedigree dogs look similar?

4 A mongrel dog is one that has been cross-bred with another type of dog. Suggest why mongrels may not be popular with dog breeders.

5 Explain why breeding pedigree dogs is an example of artificial selection.

Are pedigree dogs healthy?

When line breeding is used there is a chance that both dogs have weaknesses that are not always shown and both dogs can pass on the feature to the offspring. The chance of this occurring is increased if the breeder uses **inbreeding** (breeding with a very close relative, for example brother and sister). Problems that arise as a result of inbreeding are:

- blindness
- deafness
- heart disease
- breathing problems
- joint problems.

6 What is the difference between inbreeding and line breeding?

7 Why is there a greater chance of recessive diseases being shown in inbreeding?

8 Use your understanding of breeding to give the case for or against the breeding of pedigree animals.

FIGURE 3: Bulldogs are prone to breathing problems due to inbreeding. Can you suggest how this problem could be bred out of the dogs?

... inbreeding ... line breeding ... pedigree ... selective breeding

1 A survey was carried out of a Year 9 group. These are the results:

	Number of pupils	
Characteristic	**Male**	**Female**
Brown eyes	10	12
Blue eyes	4	5
Freckles	6	4
No freckles	8	13
Left-handed	4	3
Right-handed	10	14
Height 145-155 cm	1	3
Height 156-165 cm	5	7
Height 166-175 cm	6	6
Height 176-185 cm	2	1

Study the evidence and indicate if the following statements are right, wrong or not enough information:

a More boys have brown eyes than girls.

b All the males are brown eyed and right-handed.

c A boy is the tallest member of the group.

d There are less pupils with freckles in the group than without.

2 Every cell contains genetic information about an individual.

a What structure contains a cell's genetic information?

b What is the name of the cells that carry the genetic information from the mother and father to make an embryo?

c Write the following parts of a cell in order of size, starting with the smallest.

    nucleus   gene   chromosome

3 The domestic cow has been bred from an animal similar to the wild bison.

a In what ways are the two animals different from each other?

b How has the domestic cow been bred from the bison?

c What features would a farmer want in a milking cow?

4 Michela and Malcolm have freckles. They have a child who does not have freckles.

a Copy and complete the cross below to show how this occurred. Use F for freckles and f for no freckles.

	Michela	**Malcolm**
parent:	Ff	Ff
sex cells:	F f	F f

offspring:

	F	f
F		
f		

b Which characteristic is dominant?

c What ratio of the offspring would you expect to be freckled?

d What results would you expect if Michela had not had freckles?

5 Dandelions are growing in Sven's garden. He decides to dig them up with a spade. In doing so he leaves small parts of the roots in the ground. A few weeks latter the dandelions are back.

a How could Sven get rid of the dandelions permanently?

b The dandelion roots were cut up into pieces and all the pieces re-grew. Why are the new plants that were produced called clones?

c Why did some of the new dandelion plants look the same and some different?

6 Petra has cystic fibrosis but neither of her parents have it. Cystic fibrosis is caused by the presence of two cystic fibrosis genes in her chromosome.

a Explain why her parents do not have cystic fibrosis but she does.

b One treatment is for her to use a nasal spray which has normal genes in it. The genes enter the nose cells but do not attach to the chromosomes. Suggest why the gene does not cure cystic fibrosis permanently.

c Peta studies the Internet and finds out that genes can be added to the chromosomes in the egg cell of animals to give the animals new characteristics. This is called genetic engineering. Is it acceptable to give the human egg a healthy gene? Give clear reasons to support your argument.

Topic Summary

Learning Checklist

4

☆ I can give examples of variation that is caused by the environment. page 42

☆ I can give examples of variation that are inherited. page 42

5

☆ I know the nucleus contains the genetic information. page 42

☆ I know the cause of variation is due to environmental *and* inherited factors and the interactions between them. page 42

☆ I can explain how a fertilised egg inherits information. page 48

6

☆ I can explain the difference between the nucleus, chromosomes, DNA and genes. page 43

☆ I know the difference between reliability and accuracy when taking measurements. page 45

☆ I can use data from identical twins to identify which features are inherited and which features are caused by the environment. page 47

☆ I can explain what a clone is. page 53

☆ I can explain how artificial selection is carried out. page 54

7

☆ I know the difference between accuracy and precision when taking measurements. page 45

☆ I can explain why problems are caused by inbreeding, for example in pedigree dogs. page 55

8

☆ I can use data to predict a range of outcomes. page 49

☆ I can explain the ethical factors associated with genetics and provide reasoned arguments. page 53

Topic Quiz

1 Mia has blue eyes. She is tall and has a scar on her knee.
   a) Which of the features is caused by the environment?
   b) Which of the features is caused by the environment and genes?

2 Where are chromosomes found in a cell?

3 What is a clone?

4 What is the difference between a gene and a chromosome?

5 Why would you measure the length of my middle finger to the nearest mm rather than cm?

6 Does repeating results in an experiment increase the reliability, accuracy or precision?

7 Explain how you would breed sheep that produce lots of wool.

8 What chemicals are chromosomes made of?

9 Explain the difference between 'accuracy' and 'precision'.

10 What problems can be caused by inbreeding?

True or False?

If a statement is false then rewrite it so it is correct.

1 All twins are genetically identical

2 The sex cells pass on genetic information

3 All cells in the body have chromosomes

4 A person's weight is only affected by what they eat.

5 A person's intelligence is only inherited from their parents.

6 A person's blood group is passed on from their parents.

7 The genetic information in a plant cell is found in the nucleus.

8 If I repeat my measurements I improve the accuracy.

9 Breeding pedigree dogs always produces the healthiest animals.

10 DNA is found in the nucleus.

11 Clones always form exactly the same as the parent they are taken from.

Literacy Activity

Can a man breed with an ape? Early farmers bred a horse with a donkey to produce a mule. This animal is very hardy and the gold miners used to use them to carry their heavy tools. Unfortunately the mule is infertile. Scientists have bred a lion with a tiger to produce a liger. This had to be carried out in a test tube where the sperm was injected into the egg. The animal is again infertile.

Animals created by mixing different species are called chimera. For a long time science fiction has looked at the idea of creating a half-man half-ape animal.

A mule.

1 What does 'breed' mean?
2 Give **one** advantage of breeding a mule and **one** disadvantage.
3 What problems would be caused if a half-man half-ape was bred?

ICT Activity

Design an animal that could be useful to man if two other animals were bred together. You will need to indicate its strengths and construct a diagram of what it would look like. You could compare your creation with others in your group.

Treasures of Sutton Hoo

The greatest treasure found so far in Britain belonged to an Anglo-Saxon King. He is believed to have been Raedwald of East Anglia, who died in 625 AD. The treasure was found at Sutton Hoo in Suffolk where there are lots of burial mounds arranged along a hill that overlooks the river Deben. When King Raedwald died, a ship was hauled up from the river to act as his tomb. Weapons, armour, coins and jewellery were arranged around his body and then the ship was covered with earth forming a large mound.

Treasures of an Anglo-Saxon King found when his burial site at Sutton Hoo was excavated in 1939.

BIG IDEAS

By the end of this unit you will be able to explain how metals are affected by chemical reactions and use information to develop and use the reactivity series. You will be able to represent reactions by equations and to explain reactions by using the particle model.

Why were only treasures made from gold found at Sutton Hoo?

The soil at Sutton Hoo is very acidic. The combination of acid and water in soil destroys many kinds of materials. There were rusty stains in the soil at the burial site that were all that remained of iron weapons. The leather belts and fabrics and even the body of the king were destroyed by the acidic conditions. What remained in the ship burial were mostly objects made from unreactive metals such as gold.

The golden jewellery was clearly the work of master goldsmiths. Even the fine detail of animals and plants in the designs remain clear and shiny to this day. Iron goes rusty and even copper and bronze objects slowly turn green in acidic conditions.

The designs of gold coins found at the burial site provide evidence for the date of the king's death in the 7th Century.

What do you know?

1   In which century were the treasures of Sutton Hoo buried?

2   Whose tomb was it at Sutton Hoo?

3   What is special about the soil conditions at Sutton Hoo?

4   What effect do these soil conditions have on organic materials such as leather and fabrics?

5   Name **one** metal that is easily corroded by acidic soils.

6   What evidence from a steel sword might an archaeologist expect to find in the soil?

7   Why was gold almost the only preserved metal found in the ship burial?

8   How does the presence of gold help fix a date for the burial?

9   Give **three** examples of other metals that are used nowadays to make jewellery.

10 Why might it be harder to understand an archaeological site if there were no metallic objects present?

The alkali metals

BIG IDEAS

You are learning to:
- Identify why some metals are unusual
- Interpret the pattern of reactions of alkali metals
- Use a pattern to put metals in order

All about sodium

Most metals are **dense**. Knives and forks are dense – they sink in water. They do not change when they are put in water.

Sodium metal is different. It does change when it is put in water – it **reacts**. Sodium floats in water and fizzes! It is a soft metal that can be easily cut with a knife. It also changes in air, so it is stored under oil in a bottle.

1 If something is not very dense, does it sink or float in water?

2 How do we know that sodium reacts with water just by looking?

How Science Works

Look at the picture. What evidence is there that the nail is made from metal? What extra information is given by the magnet?

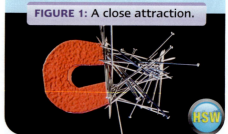

FIGURE 1: A close attraction.

The alkali metals

The table below compares the **properties** of the first three metals in **Group I** of the **Periodic Table**. The Group I metals are called the **alkali metals**.

There are many similarities between the alkali metals but they are not identical.

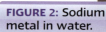

FIGURE 2: Sodium metal in water.

Alkali metal	Symbol	Does it float in water?	How is it stored?	Is a gas produced when it is in water?	How reactive is it?
Lithium	Li	yes	in oil	yes	slow bubbling
Sodium	Na	yes	in oil	yes	fast bubbling, melts
Potassium	K	yes	in oil	yes, gas ignites	very fast, violent
Rubidium	Rb	no	in oil	yes, ignites	violent
Caesium	Cs				

- Most of them react in air and in water and must be stored under oil to keep oxygen and moisture out.
- Most, not all of them, float and fizz in water. The gas bubbles contain hydrogen. Potassium is so reactive that the heat produced in the reaction sets light to the gas.

... alkali metals ... dense ... Group I ... hydroxides ... Periodic Table ...

When sodium reacts with water and the water is tested with litmus indicator it turns blue. The water contains an alkali called sodium hydroxide. The other alkali metals also react in water to produce a metal hydroxide. So we write:

sodium + water ➔ sodium hydroxide + hydrogen.

alkali metal + water ➔ alkali + hydrogen gas

This is why metals in Group I are named the alkali metals.

3 Which is the most reactive metal out of lithium, potassium and sodium?

4 Write similar word equations for lithium and for potassium with water.

Reactivity order

The reactions of the alkali metals are similar – they show a trend of becoming more reactive down the group. For example, potassium is more reactive than lithium. The **reactivity trend** is:

Trends in reactivity allow us to make predictions about reactions we have not tried.

We have seen that alkalis formed in water all have the same endings – they are **hydroxides**.

potassium + water ➔ potassium hydroxide + hydrogen

substances on the left-hand side are **reactants**

substances on the right-hand side are **products**

We can represent these reactions using chemical symbols.

$2Na + 2H_2O \rightarrow 2NaOH + H_2$ (balanced equation)

5 Write the symbol equations for lithium and for potassium with water.

6 What are the products of the reaction of rubidium in water?

7 Look back at the table in the Level 5 panel. Examine the pattern in the table. What trend can you deduce for the metal densities? Explain your answer.

8 Using the reactivity trend, predict the outcome of adding caesium to water.

9 Why do we regard the reactions of alkali metals with water as dangerous?

10 When lithium is removed from oil and added to water, the reaction often starts slowly. Explain why.

Did You Know...?

Alkalis are very corrosive materials. Sodium hydroxide has a common name of caustic soda. Caustic means 'burning' and it attacks and damages the skin and eyes. It is a strong enough alkali to strip off layers of paint from old doors and to unblock drains.

The hazard warning symbol on a bottle of household drain cleaner.

The reactivity of Group I alkali metals increases down the group.

Word and symbol equations

BIG IDEAS

You are learning to:
- Interpret word equations
- Construct symbol equations
- Recognise and interpret patterns in equations

Describing a reaction

Magnesium is a metal. If you put magnesium into sulphuric acid, it fizzes. A **reaction** takes place. Magnesium and the acid are the things we start with. They are called the **reactants**. The magnesium soon disappears. A gas is made that escapes into the air. It is **hydrogen gas**. The sulphuric acid has turned into a new material that is **dissolved** (hidden) in the water. This new material is called magnesium sulphate. It is one of the **products**, together with the hydrogen gas.

FIGURE 1: Magnesium metal reacts in acid. How do you know just by looking that a reaction takes place?

Word equations

All these changes can be shown by using a **word equation**.

potassium + water $\longrightarrow$ potassium hydroxide + hydrogen

reactants

products

The arrow in the word equation shows that the reactants have *changed* into something new, the products.

1. What do we call the materials used at the start of a chemical reaction?
2. Why do word equations need an arrow?

Using a symbol equation

When black copper oxide reacts with sulphuric acid, it produces a blue **solution** of copper sulphate. We can show this by using a word equation or a **symbol equation**.

The small numbers written on the line are called subscripts. They show how many atoms of each type are in a compound. In blue $CuSO_4$ there is one copper, one sulphur and four oxygen atoms.

copper oxide + sulphuric acid $\longrightarrow$ copper sulphate + water
CuO + H_2SO_4 $\longrightarrow$ $CuSO_4$ + H_2O

A symbol equation shows how the atoms in reactants change during a reaction. It also shows clearly when two or more reactants react in the same way. For example, zinc oxide reacts in the same way in sulphuric acid as does copper oxide.

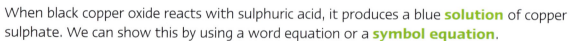
... balanced ... dissolved ... hydrogen gas ... product ... reactant

$$\text{zinc oxide} + \text{sulphuric acid} \rightarrow \text{zinc sulphate} + \text{water}$$
$$\text{ZnO} + \text{H}_2\text{SO}_4 \rightarrow \text{ZnSO}_4 + \text{H}_2\text{O}$$

3 What is the chemical formula for sulphuric acid?

4 What colour change is seen when copper oxide reacts with sulphuric acid?

5 Write the symbol equation for the reaction of magnesium oxide (MgO) with sulphuric acid.

6 Looking at the symbol equation for copper oxide and sulphuric acid, which groups of particles [groups of atoms] seem to be present on both sides?

7 What happens to the oxygen particle in zinc oxide during the reaction?

Patterns in equations

There are rules for writing symbol equations.

> *1* The numbers of atoms on each side of the arrow must be the same. Atoms cannot be created or destroyed in reactions.
>
> *2* Checking the symbol equations above shows that they are **balanced** equations: the numbers are the same on each side of the arrow.

$$\text{Mg} + \text{H}_2\text{SO}_4 \rightarrow \text{MgSO}_4 + \text{H}_2$$
$$\text{Zn} + \text{H}_2\text{SO}_4 \rightarrow \text{ZnSO}_4 + \text{H}_2$$
$$\text{Fe} + \text{H}_2\text{SO}_4 \rightarrow \text{FeSO}_4 + \text{H}_2$$

Equations are used to represent chemical reactions that have been carried out and observed.

Many metals react with acids to form a new compound, called a **salt**, and hydrogen gas.

These equations show a clear pattern. The pattern can be used to predict reactions of other metals with acids but all the reactions need to be tried to see if the predictions are correct.

8 Change the symbol equation of Fe reacting with H_2SO_4 shown above into a word equation.

9 Which gaseous product is often produced when a metal reacts with an acid?

10 Write down the symbol equation for the reaction between calcium and sulphuric acid.

11 Calcium sulphate is an insoluble compound. How does this affect the reaction of calcium metal with sulphuric acid?

How Science Works

Demonstration practical
Look at how these metal oxides react in water and in sulphuric acid. Use black copper oxide CuO and zinc oxide ZnO.

Method
1 Add a very small amount of each oxide to half a tube of water and shake.
2 Repeat using acid instead of water.
3 Leave the tubes in a beaker of hot water for 15 minutes
4 Observe any changes.
5 Tabulate your observations.

How Science Works

Experimental work is the foundation of chemical research. Scientists can look at sets of reactions and make predictions about new reactions. They verify their predictions by trying the reactions.

For example, a perfect balanced equation can be written for gold reacting with sulphuric acid but unfortunately this reaction does not actually work!

Reacting metals

BIG IDEAS

You are learning to:
- Recognise that calcium reacts quickly
- Identify similarities between calcium and magnesium
- Construct equations for their reactions

Calcium and magnesium

Look at samples of calcium and magnesium burning in oxygen and reacting with water and complete a table like this one.

Symbol	With oxygen	Metal with cold water
Mg		
Ca		

Group trends

FIGURE 2: Groups I and II.

Li 3	Be 4
Na 11	Mg 12
K 19	Ca 20
Rb 37	Sr 38
Cs 55	Ba 56
Fr 87	Ra 88

Magnesium and calcium are in **Group II** in the Periodic Table. **Elements** in the same group react in similar ways – they show trends. Magnesium and calcium burn vigorously in air. The word equation for the reaction of calcium burning in air is:

> calcium + oxygen → calcium oxide

$$2\,Ca + O\,O \rightarrow 2\,Ca\,O$$

Calcium burns with an orange-red flame. Calcium oxide (CaO) is produced, which is a white powder. The common name for calcium oxide when it is made from limestone is **quicklime**.

Magnesium also reacts violently with oxygen, burning with a brilliant white flame to leave white magnesium oxide (MgO).

Both metals react with water giving off hydrogen gas but calcium is more reactive.

Magnesium does react slowly with cold water, but it takes a few days to produce just a few bubbles of hydrogen gas.

> calcium + water → calcium hydroxide + hydrogen
> magnesium + steam → magnesium oxide + hydrogen

How Science Works

Many of the hills and mountains in Europe are made from two rocks, limestone and dolomite. They are both sedimentary rocks, often containing marine fossils. Samples of the rocks fizz with dilute acids, releasing carbon dioxide gas. Limestone is mostly calcium carbonate and dolomite is a combination of calcium and magnesium carbonates.

1 Give **three** similarities between dolomite and limestone.

2 What evidence is there that they formed beneath the sea?

FIGURE 1: Mountains from metals.

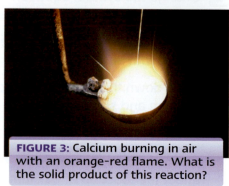

FIGURE 3: Calcium burning in air with an orange-red flame. What is the solid product of this reaction?

1 Why is it not surprising that magnesium and calcium react in a similar way, when you study the Periodic Table?

2 What is the reactivity trend in Group II?

Symbolic changes

Although the reactions of magnesium and calcium with air and water are similar, they are not identical. With air, or with oxygen alone, the equations are:

With water or steam in the case of magnesium:

$$2Mg + O_2 \rightarrow 2MgO$$
$$2Ca + O_2 \rightarrow 2CaO$$

$$Mg_{(s)} + H_2O_{(g)} \rightarrow MgO_{(s)} + H_{2(g)}$$
$$Ca_{(s)} + 2H_2O_{(l)} \rightarrow Ca(OH)_{2(aq)} + H_{2(g)}$$

Mg + O·H·H → Mg·O + H·H

The letters in brackets tell us about the water in each case. Steam is a gas so $H_2O_{(g)}$ is used and water is a liquid and is written $H_2O_{(l)}$. These labels are called **state symbols**. They are used in equations to show what **physical state** a substance is in. The symbol for solid is (s) and for solution (aq).

3 **a** The element strontium (Sr) is below calcium in Group II. Predict how fast strontium reacts with water compared to calcium.
b Write both word and symbol equations for the reaction of strontium (Sr) with water.

4 If the chemical formula for calcium sulphate is $CaSO_4$, what is the formula for strontium sulphate?

What metal compound is in this firework?

All new element

Read the details of this unknown new element and some of its properties.

Name	Symbol
Newium	Nm
How stored	In dry air
Appearance	Grey, shiny on a clean surface
With water	Sinks, bubbles, flammable gas, solution has a high pH value

5 What evidence is there that Nm is metallic?

6 Would you place it in Group I or II? Explain your answer.

7 Write a symbol equation for the probable reaction of Nm with water.

8 What is the grey coating on the surface of Nm and what happens to a weighed sample of pure Nm as the coating forms?

... physical state ... quicklime ... state symbol

Metals and acids

BIG IDEAS

You are learning to:
- Describe what you see when metals react with acids
- Construct equations for the reactions
- Use evidence from experiments to make predictions

Reaction speeds

If a metal reacts with water it is likely to react even faster with acids. Calcium fizzes more violently in acid than in water. Some metals do not react at all. Gold stays shiny in water and in acid too. Gold is very **unreactive**.

Unreactive metals are used to make jewellery and coins and for use outside in the weather.

1 Why does a gold bottle not corrode when it is filled with acid?

2 Gold never wears out so why is it not used for roofing?

Comparing reactivity

Many metals react with acids and release bubbles of hydrogen gas. We can use the **rate** of bubbling to compare the **reactivity** of metals. **Reactive** metals bubble faster than unreactive metals. This allows a **reactivity series** to be drawn up of the metals that are tested.

3 Describe an investigation to find out which of two metals is the more reactive.

4 Describe an investigation to find out how much gas is released when a piece of metal reacts with an acid.

Developing a reactivity series

We can test metal samples with a variety of reagents such as oxygen, water, acids and alkalis. Comparison of the reactivities and the products of the reactions allows us to construct a reactivity series but there are problems. Aluminium appears unreactive, we use it for window frames and tube trains. The metal is very reactive indeed but it becomes coated in an unreactive layer of aluminium oxide protects the metal underneath.

5 Iron goes rusty by reaction with oxygen and water. Why doesn't this rust coating stop any further corrosion?

Did You Know...?

Artists use acids and heat treatment to produce interesting colour effects. Some metals such as titanium can show a range of colours and surface patterns. Each piece made by the artist is unique. The technique allows artists to be very creative.

A

B

FIGURE 1: Which metal, **A** or **B**, is the more reactive?

Reactivity series

Scientists find it very useful to place metals in order of their reactivity. By doing this they can predict which metals will be useful for different jobs.

The reactivity of metals can be compared by placing the same-sized samples of different metals in the same type and volume of acid. By observing the numbers of bubbles produced, the metals can be placed in order of their reactivity.

Your teacher will provide you with the apparatus that you may need for your investigation.

Method:

1 Put four test tubes in a rack.

2 Add 2 cm^3 of dilute sulphuric acid to each tube.

3 Add clean pieces of the four metals, one in each tube. Label each tube.

4 Observe and record the numbers of bubbles released in each test tube in a table.

FIGURE 1: Gold is used to make jewellery. Why is this?

How Science Works

Planning exercise

How could you find out if the reactivity series of Mg, Zn, Fe and Cu with sulphuric acid resembles those with different acids? These might include hydrochloric and phosphoric acids.

Use similar pieces of the separate metals and test with solutions of the different acids. Control the concentrations and temperatures of the acids. Compare the reactions, for example in terms of effervescence.

When particles of an acid collide with metal particles, a reaction may occur. In terms of particles, how would reactive and unreactive metals be different?

Questions

1 Which was the most reactive metal?

2 What was the order of reactivity, starting with the most reactive metal? Write your answer using symbols.

3 Where would you place gold (Au) in your series?

Copper mining

An open-pit copper mine in the USA.

Things have changed since the Bronze Age thousands of years ago – then people found actual pieces of copper, often washed down by streams. This was native copper, just the metal itself. The least reactive metals were discovered first, for example copper, silver and gold. These three metals are all placed in the centre part of the Periodic Table. Bronze Age people melted it with tin metal to form bronze. They used the bronze to make weapons, axes and jewellery. Today, there is very little native copper left. It has to be extracted from its ores (compounds of copper) that occur naturally in the Earth's crust. In metallic ores, copper particles are combined with other elements such as oxygen or sulphur. The copper particles must be chemically separated to give the metal itself. We call this smelting.

- Copper miners working 100 years ago in Cornwall could expect to dig up ores that contained up to 10% copper. Today miners find ores that contain less than 0.5% metal – this is all that is left.

- With a metal ore that is 99.5% waste, mining needs to be carried out on a huge scale to produce enough copper for wiring and plumbing needs. Copper and lead have some very useful metallic properties. They are malleable, conductors and easy to alloy with other metals to improve their properties.

- Copper is often extracted from its ore by soaking it in sulphuric acid. This process is called leaching. It gives a blue solution containing copper sulphate.

As metal ores become scarcer and more and more expensive we need to find ways to make supplies of essential metals last. Two possibilities are:

- metal recycling
- the use of alternative materials.

Brass is an alloy of copper and zinc. Almost all of the brass used today is recycled material. This reduces the need for more mines producing new copper and zinc. In plumbing, modern houses often use plastic pipes as an alternative to copper. This preserves limited supplies of the metal for essential uses such as electrical wiring.

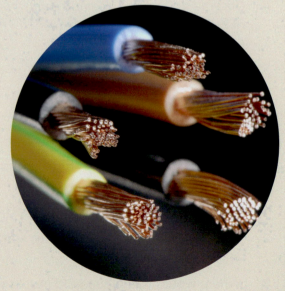

Copper wiring used in electric circuits.

Assess Yourself

1 Which **two** metals are used to make bronze?

2 Give **three** uses of bronze.

3 What are natural compounds of metals called?

4 How much metal could you extract from 1000 g of an ore containing 1% metal?

5 Why are copper mines operated on such a large scale today?

6 Why is brass a good example of conservation?

7 Give **two** ways in which supplies of metal ores can be conserved for the future.

8 What are the environmental gains of recycling metals?

History Activity

Research the uses of bronze in the Ancient World. You might consider bronze used in coins or jewellery or for statues and why it was such a good material to use.

Geography Activity

Find out where in the world the top **three** major producers of a particular metal are located. You could choose from copper, aluminium or iron.

Level Booster

8 Your answers demonstrate an extensive knowledge and understanding of the properties of materials and the Earth.

7 Your answers show an advanced understanding of the wide range of processes related to the Earth and the properties of materials.

6 Your answers show that you can describe processes that relate to materials and the Earth and that you can use appropriate terminology

5 Your answers show that you can describe processes related to materials and the Earth, using more than one step or that you can use a model to explain.

4 Your answers show that you can describe some processes related to the Earth and materials and that you understand the importance of evidence.

Predicting reactions

BIG IDEAS

You are learning to:
- Recognise that predictions can be useful
- Explain how to use data to predict reactions **HSW**
- Construct equations to match your predictions **HSW**

Predicting displacement reactions

In chemical reactions we can often make predictions. If we have seen several reactions that are similar, we can **predict** other reactions we have not tried. If four different metals all fizz with acids we might predict that another metal would fizz too.

By testing many metals to see how they react with oxygen, water or acids, we can develop a **reactivity series**. The most reactive metals are written at the top with the metals becoming less reactive down the list. More reactive metals **displace** less reactive ones. The reactivity series is used to predict which displacement reactions will be successful.

Symbol	Name	Reacts with oxygen?	Reacts with water?	Reacts with dilute acid?
K	potassium	✔	✔, fast in cold	✔
Na	sodium	✔	✔, fast in cold	✔
Ca	calcium	✔	✔, fast in cold	✔
Mg	magnesium	✔	✔, slow in cold, fast in steam	✔
Al	aluminium	✔	✔, needs steam	✔
Zn	zinc	✔	✔, fast in steam	✔
Fe	iron	✔	✔, fast in steam	✔
Pb	lead	✔	no	✔
Cu	copper	✔	no	no
Au	gold	no	no	no

Does iron displace copper?

The table above shows that iron is more reactive than copper. Therefore this displacement reaction can be predicted to be successful:

iron + copper sulphate ➔ copper + iron sulphate ✔

... aluminium oxide ... displace

Does copper displace iron?

The reaction between copper and iron sulphate can be predicted not to be successful according to the data.

copper + iron sulphate → iron + copper sulphate ✗

1. a Which metal in the table do you predict could not displace any other metal?
   b Name one metal that could displace magnesium from magnesium sulphate solution but would be too dangerous to try. Explain why.
   c Why should the surfaces of metals be cleaned before testing displacement reactions?
   d Write both word and symbol equations for the reaction of zinc with iron sulphate, $FeSO_4$.

Describing displacement reactions

With blue copper sulphate it is obvious if a displacement reaction has occurred because there is a colour change. Most solutions are colourless and there may not be a colour change. Often there is a temperature change and this shows that a reaction has occurred.

Balanced symbol equations can be written – even for reactions that do not work.

$$Zn + MgSO_4 \rightarrow ZnSO_4 + Mg$$
$$Mg + ZnSO_4 \rightarrow MgSO_4 + Zn$$

Only one of the reactions shown by the equations above is predicted to work. It is the reaction between magnesium (Mg) and zinc sulphate ($ZnSO_4$). From the data in the table on page 76 magnesium is more reactive and displaces zinc.

2. Write a word equation for the reaction between zinc and lead nitrate solution.

3. Write a symbol equation for the reaction between magnesium and iron sulphate.

4. Predict if the reactions in **Q2** and **Q3** will be successful.

Particles and bonds

The number of metal particles that can combine with sulphate particles can vary. Metals can have different numbers of bonds to combine with other groups or elements. For example:

Sodium sulphate Na_2SO_4

Zinc sulphate $ZnSO_4$

Aluminium sulphate $Al_2(SO_4)_3$

We need to find out the numbers of bonds before writing more difficult chemical formulae. The sulphate particle in the examples has two bonds.

Corrosion of metals

What is rust?

Cars are made from **steel**. Steel is mostly made from the **element** iron. The problem with iron is that it **rusts**. If the paint chips off a car, the iron underneath starts to go rusty. Rust can even spread under paint. Rusty iron is brown in colour. Iron loses its strength as it goes rusty. This is why rusty cars become unsafe to drive and end up in a scrap yard.

1 Which element that is in steel goes rusty?

2 Describe **two** changes to iron caused by rusting.

FIGURE 1: Rusty cars end up in scrap yards. Why are rusted cars unsafe to drive?

The golden alternative

Rusting is just one example of **corrosion**. When some metals react with air and water they start to corrode. They lose their shiny appearance and become dull. In some cases the corrosion continues until no metal remains, as with iron.

FIGURE 2: Why does gold jewellery stay shiny whereas copper jewellery eventually turns green?

But some metals resist corrosion. They are **unreactive** metals that are not affected by air or water, or metals that react extremely slowly. Examples of unreactive metals are gold and platinum. A golden car would never corrode but it would be too expensive and also very heavy! Copper and lead corrode slowly but are still used for roofing.

Copper jewellery also corrodes slowly – it turns green. It can even stain your skin green too.

Aluminium is the shiny metal we use for cooking foil. It does not seem to corrode, even at high temperatures inside an oven. Aluminium is a special case. The metal reacts rapidly with oxygen in the air to form aluminium oxide which protects the metal underneath from further corrosion.

FIGURE 3: St Paul's Cathedral in London. What is its dome made from?

... alloyed ... corrosion ... element ... rust

3 Give **two** examples of metals that resist corrosion.

4 What colour change is seen when copper corrodes?

5 Suggest how acid rain affects the speed at which metals corrode.

Making the right choice

Metals that resist corrosion have special uses. Gold, silver and platinum are all used to make jewellery, coins and medals. It is important that metals chosen for these objects do not corrode, for example iron coins would be unsuitable. Fine gold wires are used in computers. If these wires corroded, data would be lost or the computer might stop working.

Iron can be made to resist corrosion if it is **alloyed** (mixed) with chromium and nickel which gives **stainless steel**. This shiny alloy is used for cutlery and for the insides of washing machines and in hospitals where corroded metals might harbour germs.

6 Name **two** metals that are used in coinage.

7 Which elements are alloyed with iron in stainless steel?

8 **a** Why are the insides of dishwashers made from stainless steel?

b Where in the reactivity series would you find metals such as gold and platinum?

c Why are you unlikely to read equations that show gold displacing another metal?

d Write a word and symbol equation for the reaction of sodium with silver nitrate solution, $AgNO_3$ solution. What problems might you find in testing this reaction?

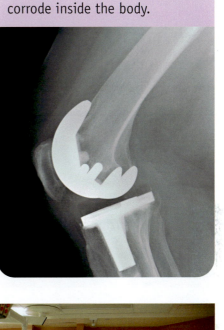

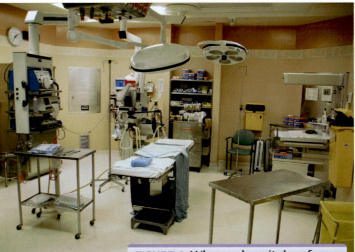

FIGURE 4: Why are hospital surfaces often made from stainless steel?

Why use steel?

Both iron and steel go rusty and yet we still use steel to make millions of cars each year. There are several reasons for choosing steel. Steel is a low-cost construction material compared with alternatives such as aluminium or stainless steel. Steel is malleable and very strong. Both galvanising (zinc coating) and painting can slow down the rusting process.

9 If the nitrate particle has a single bond to attach to other things, how many bonds must be formed by aluminium and by silver in this equation?

$$Al + 3AgNO_3 \rightarrow Al(NO_3)_3 + 3Ag$$

How to stop corrosion

Keeping iron dry

The things that make iron rust are:

- water
- air (oxygen).

FIGURE 1: These have not rusted because the climate is dry.

Dry iron does not go rusty. In a desert there is very little water and so iron hardly rusts at all.

In our climate we need ways of keeping iron dry to stop it from rusting.

- Iron can be painted, for example cars and trucks.
- Iron can be covered in plastic, for example a washing line or clothes-drying rack.
- Iron can be covered in grease or oil, for example a bicycle has oil or grease put on its iron chain.

All of these methods work by keeping water away from iron.

1 Why do cars not go rusty in deserts?

2 **a** How can you stop a bicycle chain from rusting?

b Water cannot soak through plastic or grease. Explain why this helps keep bicycle chains shiny and plastic-coated steel washing lines in new condition.

Zinc to the rescue

Iron and steel can be coated with zinc to stop the iron underneath **corroding**. The process is called **galvanising** and the product is **galvanised iron**.

Did You Know...?

The steel cables supporting the Forth bridge are slowly corroding and some thin wires have already snapped.
Engineers are pumping a current of dry air through the cables to stop the corrosion. Without this, the bridge might collapse.

Forth road suspension bridge, near Edinburgh in Scotland.

Did You Know...?

In the 19th Century much of the centre of Paris was rebuilt. The architect chose zinc sheeting as the roofing material. Zinc corrodes very slowly and produces an attractive grey surface, called a **patina**. These zinc roofs can last for up to 100 years and have been favourite subjects for artists in Paris to paint.

... anodising ... corroding ... galvanised iron

For iron to rust oxygen and water must be present. If either oxygen or water is excluded, the iron does not corrode. The zinc coating excludes oxygen and water.

An example of galvanising is in iron roofing sheets and nails. They are coated with zinc to protect them from the weather. (Zinc is too weak to make nails.)

Food cans are often called 'tins' because the steel can is covered with a thin layer of tin. This prevents the steel from corroding and possibly reacting with the food inside the can. Zinc cannot be used for cans because some foods dissolve the zinc coating.

FIGURE 2: Galvanised iron roofing sheets. Why is it necessary to galvanise iron materials that are used outside?

FIGURE 3: Why is aluminium anodised? Why do anodised aluminium objects look attractive?

3 What is the corrosion coating called that forms on some metals?

4 Tin follows iron in the reactivity series. Why is this a problem when a tin-plated steel can is scratched?

5 The sunken wreck of the ship *The Titanic* is surrounded by cold seawater. Why has the wreck not rusted away completely?

Stopping the rot

Aluminium does not corrode because it has a coating of aluminium oxide on its surface that is unreactive. This coating can be improved by **anodising** the metal. During this process an aluminium object is placed in a special electrical cell that produces oxygen and this strengthens the surface coating. The coating can absorb bright colours. We see examples of this in Christmas decorations and in cookware.

Most cars are now partly galvanised to resist corrosion. Many layers of paint are sprayed on to further protect the steel underneath. Some modern cars are made completely from aluminium, just like aircraft and underground trains. The aluminium does not need to be painted. Stainless steel planes would be too heavy even to take off.

6 Give **two** ways in which cars can be protected from corrosion.

7 Why might you describe a steel plane as an 'expensive caravan'?

Designer problems

Aluminium cars do not go rusty, unlike steel. The reactive aluminium is protected by an impermeable surface layer of aluminium oxide that protects the metal. However, aluminium is much more expensive than steel to produce. We could also use the unreactive metal titanium to build cars but they would cost a fortune to buy.

8 Aluminium requires large amounts of expensive electricity to extract from bauxite ore. Why is this an argument in favour of recycling drinks cans?

... galvanising ... patina

1 For each of the following statements write 'T' if the statement is true or 'F' if it is false.

   a Iron needs water and air to rust.

   b Rusting is not an example of corrosion.

   c Metal ores are natural compounds.

   d Gold coins never rust.

2 Copy and complete the sentences using the words below.

       equations   gold   more   series

   a Some metals are _____ reactive than others.

   b A reactivity _____ shows how the properties of metals compare.

   c One metal that is found as a pure metal is _____ .

   d We can show what happens in a reaction using _____ .

3 Write down the correct description for each word.

displacement	corrosion of iron
dilution	speed of reaction
rusting	take the place of
rate	adding water to a solution

4 Write down the correct metal for each use.

Metal:	Use:
gold	drinks can
aluminium	steel in cars
copper	jewellery
iron	electrical wiring

5 Grey zinc powder is added to a blue solution of copper sulphate. The solid changes colour to brown and the solution becomes colourless. For each of the following statements write 'T' if the statement is true or 'F' if it is false.

   a Zinc has displaced copper.

   b The final solution contains zinc sulphate.

   c Zinc is less reactive than copper.

   d A finer zinc powder would react more slowly.

6 Copy and complete the equations below.

a sodium + water ➤ sodium hydroxide + _____

b iron + _____ + air ➤ rust

c magnesium + copper sulphate ➤ _____ + magnesium sulphate

d aluminium + iron oxide ➤ iron + aluminium _____

7 Use the following reactivity series to write 'T' if a statement below is true or 'F' if it is false.

K   Na   Ca   Mg   Zn   Fe   Pb   Cu   Au

a The most reactive metal is potassium.

b Magnesium can displace lead from solution.

c The reactivities of zinc and of iron are similar.

d The metal least likely to be found as a native element is gold (Au).

8 a Why do gold objects usually provide more detailed information to archaeologists than iron objects?

b What is special about the soil at Sutton Hoo that affected the preservation of the burial goods?

c Why do we extract copper from ores that contain only 0.5% copper minerals?

d Why are copper mines so big?

9 Copy and complete and balance the following equations.

a $Na + H_2O$ ➤ $NaOH$ + _____

b $K + H_2O$ ➤ _____ + _____

c $Mg + O_2$ ➤ MgO

d Ca + _____ ➤ CaO

10 Aluminium is found towards the top of the reactivity series of metals.

a What is formed when aluminium reacts with oxygen?

b What is the formula of the product?

c Why does the aluminium in a drinks can not react with the contents?

Learning Checklist

☆ I know that some groups of metals have special names.   page 62

☆ I know that iron rusts.   page 78

☆ I know that acids react with metals.   page 64

☆ I know that gold is unreactive.   page 68

☆ I know what 'concentrated' and 'dilute' mean.   page 72

☆ I know how the alkali metals react with water.   page 62

☆ I know that rusting is an example of corrosion.   page 78

☆ I know that metals react at different rates with acids.   page 68

☆ I know why gold is chosen to make jewellery and coins.   page 78

☆ I know how to vary the concentration of an acid.   page 72

☆ I know that the Earth's supply of metals is limited.   page 78

☆ I know the reactivity trend for the alkali metals.   page 63

☆ I know that reactivity is linked to rate of corrosion.   page 78

☆ I can draw up a reactivity series of metals.   page 68

☆ I can decide which metal to use based on reactivity.   page 79

☆ I can predict the outcomes of displacement reactions.   pages 74–75

☆ I can write and interpret word equations.   page 64

☆ I can predict the reactivity of the alkali metals from data.   page 63

☆ I can understand how to prevent metals corroding.   page 79

☆ I can interpret the behaviour of metals in terms of the reactivity series.   page 69

☆ I can understand the importance of gold in archaeology.   pages 60-61

☆ I can write symbol equations for displacement reactions.   page 75

☆ I can assess the importance of recycling in the conservation of metals.   page 81

☆ I can use both scientific and economic data to decide on how best to use metals   page 81

☆ I can interpret formulae and equations in terms of the numbers of bonds used by the particles involved.   page 77

Topic Quiz

1 Which metal goes rusty?

2 Name **one** yellow metal that does not corrode.

3 Why does a new car not go rusty when left in the rain?

4 How can you dilute an acid?

5 What do you see when magnesium reacts with an acid?

6 Which gas is produced when reactive metals are placed in an acid?

7 Which group of metals includes sodium (Na) and potassium (K)?

8 What happens in a displacement reaction?

9 What does 'reactivity series' mean?

10 How much copper ore is contained in the material mined from most copper mines?

   A 5%   **B** 0.5%   **C** 2.5%?

11 What happens when copper ore is leached?

12 What is the link between acid concentration and rate of reaction with metals?

True or False?

If a statement is false then rewrite it so it is correct.

1 Alkali metals include sodium and potassium.

2 Alkali metals are not very reactive.

3 Rusting is an example of corrosion.

4 Iron needs only water to rust.

5 Magnesium is more reactive than copper.

6 Copper can displace magnesium from solution.

7 An acid can be made more concentrated by adding water.

8 A reactivity series allows reactions to be predicted.

9 Symbol equations give us more information than word equations.

10 A low-grade copper ore contains less than 0.5% copper.

11 Scrap iron can be used to recover copper from solutions containing copper sulphate.

12 The compound formed in the reaction between potassium and water is KOH.

Literacy Activity

Write the text for a poster urging people to recycle aluminium drinks cans. Make it clear what the environmental benefits would be.

ICT Activity

Use the British Museum website to research the range of metal objects discovered at the Sutton Hoo ship burial site. Display your results to show examples and write a brief description of each one.

White gold of France

Anyone who has been swimming in the sea knows that it tastes very salty. There are many different salts dissolved in sea water, including potassium salts, but the main one is sodium chloride, common salt. People have obtained salt by the evaporation of sea water for thousands of years. In Britain, there are the remains of Roman salt works as far north as Edinburgh. The salt makers of Brittany in northern France established themselves in prehistoric times. The whole landscape of the salt marshes is now divided up into a network of salt pans. Sea water is allowed in to fill the shallow pools and then left to dry up. As water evaporates, the remaining solution becomes more saline until salt starts to crystallise. The Breton salt makers said that salt was the child of the sun and the wind. Salt makers use rakes to scrape the glistening white crystals into heaps to dry before the salt is sold.

What do you know?

1 What evidence is there that the sea is salty?

2 Name a salt that is found in sea water.

3 How do we know that the Romans produced salt in Britain?

4 What physical change turns sea water into salt?

5 What does it mean to say that salt is the child of the sun and the wind?

6 What modern word is derived from *salarium*, the word the Romans used to describe their part-payment in salt to their soldiers?

7 What do we mean by a saline solution?

8 How is salt used in the food industry?

9 How might global warming affect the location of sea salt production?

10 What health risk is associated with a high intake of sodium chloride?

FIGURE 1: Sea salt called white gold.

Before the days of fridges, salt was one of the few ways of preserving fresh food. Both salt fish and salt beef are still made today. In the 18th century the British navy relied on salted meat for its crews. The Breton fishing industry brought in herrings and one tonne of salt was needed to preserve four tonnes of fish. The high concentration of sodium chloride prevented bacteria from growing and spoiling the food.

Metals and non-metals

BIG IDEAS

You are learning to:
- Describe the distinctive features of metals
- Recognise exceptions to patterns **HSW**
- Provide evidence for links between properties and uses **HSW**

Spot the metals

Metals and non-metals are different kinds of chemical elements. The special properties of metals make them easy to recognise. Metals are usually shiny, **malleable** and strong. They also conduct electricity and heat very well. Iron, gold and copper are like this. In the Periodic Table we find the metals on the left-hand side. Most of the elements are metals. Non-metals are different. Some are gases such as oxygen, nitrogen and hydrogen. Others are solids such as sulphur, the yellow element produced by volcanoes. If you could compare identical cubes made of the elements, the metals are heavier for the same size, they are more **dense**.

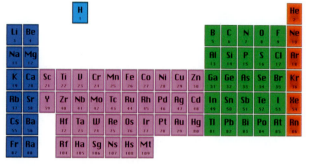

FIGURE 1: The Periodic Table

1. Give **two** special properties of metals.

2. How are the densities of metals and non-metals different?

Exceptional elements

The descriptions of the properties of metals and non-metals are useful generalisations. However, there are exceptions such as the liquid metal called mercury and the alkali metals whose densities are so low they float on water. The non-metal graphite, one form of carbon, conducts electricity but the other form, diamond, does not. Diamond and graphite are the allotropes of carbon. Although both contain the same particles, carbon atoms, the particles are arranged differently. The graphite structure has a weak point, the bonds in between layers of carbon particles. Graphite is soft enough to mark paper.

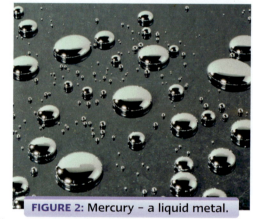

FIGURE 2: Mercury – a liquid metal.

There is a continuous change across the Periodic Table (see back of book) from metallic properties on the left to non-metallic ones on the right. There is a clear pattern in the table from the metallic properties of groups 1 and 2 across to the non-metallic ones of later groups such as oxygen in group 6 or chlorine in group 7. An element must possess several of the characteristic properties before we can be sure it really is a metal or a non-metal. There are some **semi-metals** such as silicon and germanium whose properties seem to be in-between.

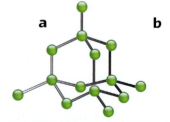
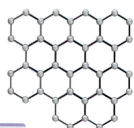

FIGURE 3: The structure of diamond (a) and graphite (b).

The special properties of silicon explain its use in making integrated circuits for computers and other digital devices.

3 What is unusual about the conductivity of carbon?

4 What category of elements includes germanium?

5 Can you identify an element as a metal if it conducts electricity?

FIGURE 4: The semi-metal silicon.

Science in the car

There are many different designs of car alarm. One simple version uses a mercury tilt switch. The metal mercury is a liquid at normal temperatures, it only freezes solid at minus 39°C. The switch contains a tube with a small amount of mercury and two electrical contacts. If the car is moved, the mercury flows along the tube completing the circuit and triggering the alarm. This is a perfect match of the properties of the metal with the way it is used.

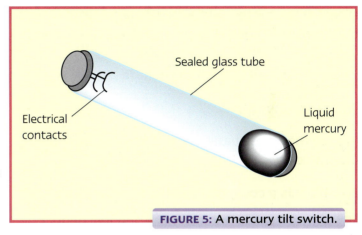

Sealed glass tube

Electrical contacts

Liquid mercury

FIGURE 5: A mercury tilt switch.

6 Give **one** use of the carbon allotrope called diamond that matches its properties.

7 Why does a mercury tilt switch still work in cold weather in Britain?

8 An element is grey, has a density of 0.5 g/cm^3, conducts electricity and heat. What category of element is it?

Did You Know...?

Modern low-energy light bulbs produce much more light than the old filament lamps using the same amount of electricity. Unfortunately they contain some toxic mercury vapour which may cause a pollution problem when old lamps are disposed of.

Metallic conductors

In metals the particles can release outer electrons that are mobile and can move freely. This accounts for the high conductivities of metals. In non-metals such as solid sulphur, all the electrons are firmly held in place and so non-metals are non-conductors (insulators).

9 What can you conclude about the electrons in the non-metal graphite which is able to conduct electricity?

10 How do the properties of metals and non-metals explain the design of a three-pin plug?

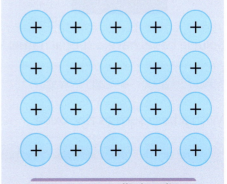

FIGURE 6: Metallic bonding.

Acids, alkalis and bases

Chemistry in the stomach

Your stomach produces acid to help digest your food. If there is too much acid, it can be painful. Your stomach is upset. We can use neutralisation reactions to make the stomach feel better. Seltzer tablets and Milk of Magnesia both contain a **base** that can neutralise the stomach acid.

Acid + base → a new compound (a salt) + water

They work like this:

We say that these bases are **antacids** because they neutralise acids.

1 What is a common cause of an upset stomach?

2 Why do seltzer tablets make you feel better?

FIGURE 1: Neutralising an acid.

Bases and alkalis in neutralisations

FIGURE 2: Alkalis and a base.

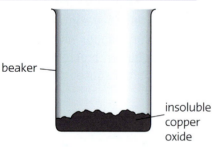

beaker

insoluble copper oxide

Sodium hydroxide

Potassium hydroxide

Soluble bases have a special name. We call them alkalis. This produces a similar general reaction:

Acid + alkali → salt + water

With a base such as zinc oxide:

nitric acid + zinc oxide → zinc nitrate + water

... alkali ... antacid

The neutralisation reaction can be monitored using colour change indicators or a pH meter. The low pH of the acid solution rises to pH7 when it has been neutralised and a salt has formed.

3 Copper oxide is insoluble in water. Is it a base or an alkali?

4 Complete the word equation:

hydrochloric acid + zinc hydroxide ⟶

5 What colour change would you predict for universal indicator in the reaction in question **4**?

Particle patterns

During neutralisation, the particles are rearranged to form new products. For example, the hydrogen of an acid may combine with the oxygen from a base to form water.

$$H_2SO_4 + ZnO \longrightarrow ZnSO_4 + H_2O$$

Formulae and equations

If you know the **chemical formulae** of the reactants and products, you can write balanced equations for neutralisations. In a balanced equation the total number of each sort of atom is the same on each side.

Acid	Alkali	Salt	Water
a) H_2SO_4	NaOH	Na_2SO_4	H_2O
b) HCl	KOH	KCl	H_2O

The balanced equations look like this:

a $H_2SO_4 + 2 NaOH \longrightarrow Na_2SO_4 + 2 H_2O$
b $HCl + KOH \longrightarrow KCl + H_2O$

In both cases the pattern is the same. A hydrogen particle (H) from the acid combines with a hydroxide particle (OH) from the alkali to give water.

6 What is the formula of sulphuric acid?

7 Which particles are present in all alkalis?

8 How many hydroxide particles would be needed to neutralise one particle of phosphoric acid H_3PO_4? Explain your answer.

9 Write a balanced symbol equation for the neutralisation of phosphoric acid by potassium hydroxide.

Did You Know...?

Copper is not a very reactive metal and so we can use it outside for roofs or statues. It does weather very slowly and becomes covered in a layer of green. This green compound protects the metal underneath from any further corrosion. It is a basic compound and could react with acids to neutralise them.

Carbonates and acids

Get fizzing

Some rocks are made of calcium carbonate. These include chalk, limestone and marble. Even sea shells contain the same chemical. **Geologists** are scientists who study rocks. They use the fizzing reaction with acid to identify these rocks. Other kinds of rock may look the same but they don't fizz with acids. The gas released is invisible. It is the same one you find in fizzy drinks and it is called carbon dioxide. Geologists usually carry small plastic bottles of hydrochloric acid when out testing carbonate rocks.

1. Why do sea shells fizz with acids?

2. Why do geologists prefer to use plastic bottles to carry acid rather than glass bottles?

FIGURE 1: Testing carbonate rocks.

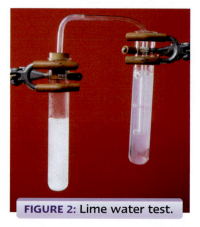

FIGURE 2: Lime water test.

Evidence for a reaction

Carbon dioxide extinguishes a burning splint, this is why the gas is used in some fire **extinguishers**. However, this is not a satisfactory chemical test for carbon dioxide since other gases, such as nitrogen, behave in the same way with flames. Lime water is a solution of the base calcium hydroxide, it is strongly alkaline with a high pH value. When carbon dioxide is bubbled through lime water it changes, as shown in Figure 2. No other gas is able to turn lime water cloudy so it is a good test for the gas.

All gases have a mass. When carbonates react with acids the gas escapes into the air. The mass of the flask shown in Figure 3 will go down as the carbon dioxide escapes. We can measure the mass at regular intervals to monitor how it changes with time.

3. How does carbon dioxide affect lime water?

4. How could you show that lime water is a solution of a base?

5. How can you tell when the reaction between a carbonate and an acid is finished?

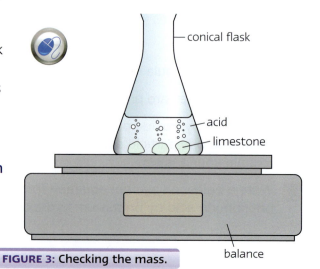

conical flask

acid

limestone

balance

FIGURE 3: Checking the mass.

extinguisher

Carbonate and acid reactions

There is a general reaction for carbonates and acids

Acid + carbonate → salt + water + carbon dioxide gas

When sulphuric acid reacts with copper carbonate or with limestone (calcium carbonate) we have:

Sulphuric acid + copper carbonate → copper sulphate + water + carbon dioxide

Sulphuric acid + calcium carbonate → calcium sulphate + water + carbon dioxide

Calcium sulphate is not very soluble in water.

6 **a** Would carbonates react faster as lumps or as powder? Explain your answer.

b How would you expect the rate of the second reaction to change, the one with limestone and sulphuric acid? Explain your answer.

Explaining the changes

Figure 4 is a graph showing how the mass changes during the reaction between calcium carbonate and hydrochloric acid. The reaction goes most rapidly at the start and the mass falls steeply. As the acid and carbonate are used up, the reaction slows down. When all of the acid or the carbonate has reacted, no more gas is produced. The mass stops changing and the line on the graph becomes horizontal.

FIGURE 4: Plotting the mass changes.

7 At what stage of the reaction is the mass changing most quickly?

8 How could we tell if it was the acid or the carbonate that initially was in excess?

9 What would the graph look like if you measured the volume of gas produced instead of the mass?

Particle pictures

In a concentrated solution of acid, the acid particles are close together. There will be lots of collisions with the surface of the solid carbonate. The number of successful collisions will be higher than in dilute acid. This explains why the reaction rate is faster with more concentrated acids. This explanation matches the shape of the graph (above). As the number of unreacted acid particles diminishes, the reaction slows down.

10 How would the graph (above) be different if you used a more concentrated acid solution with the same amount of calcium carbonate?

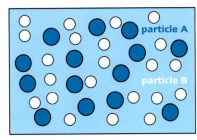

Concentrated solution, many collisions

FIGURE 5: Collisions in solution.

Dilute solution, few collisions

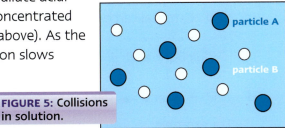

Salts

BIG IDEAS

You are learning to:
- Explain what a salt is
- Describe the general salt preparations
- Interpret the formulae of salts

HSW

Looking at salts

All acids contain hydrogen. You can see this by comparing the formulae of some acids such as hydrochloric acid, nitric acid or sulphuric acid. When acids are used in chemical reactions we can replace this hydrogen with a metal particle. The new compound is called a salt. Salts form crystals, often with beautiful colours or shapes. Some crystals are natural and found in rocks.

FIGURE 1: Crystals of salts.

1 Which chemical element do all acids contain?

2 What must replace this element to make a salt?

FIGURE 2: Natural crystals of gypsum. These crystals form naturally when water evaporates.

General reactions

A set of very similar chemical reactions can be called **general reactions**. As we have seen on pages 91 and 93 these reactions can produce new salts.

a acid + metal oxide (a base) → salt + water

An example would be:

Sulphuric acid + copper oxide → copper sulphate + water

b acid + alkali → salt + water

For example:

Sulphuric acid + sodium hydroxide → sodium sulphate + water

c acid + metal carbonate → salt + water + carbon dioxide gas

Sulphuric acid + copper carbonate → copper sulphate + water + carbon dioxide

d acid + metal → salt + hydrogen gas

For example:

Hydrochloric acid + magnesium → magnesium chloride + hydrogen

... general reaction ... subscript

All of these reactions produce new salts from the original acid. The names of salts are derived from the acids.

FIGURE 3: Calcite is calcium carbonate.

The pure salts can be recovered using evaporation. Slow evaporation gives better shaped and larger crystals since they have time to grow.

3 What do we call the salts of (a) nitric acid; (b) hydrochloric acid?

4 Name a rock that contains the same salt as calcite.

5 Why do we add an excess of copper carbonate to sulphuric acid when preparing the salt copper sulphate?

FIGURE 4: Sodium chloride.

Formulae of salts

The chemical formulae of salts tell us both the identity of the salt and the numbers of each atom present. Small numbers written down on the line are called **subscripts**. The subscript multiplies the chemical symbol written before it. $MgSO_4$ means one atom each of magnesium and sulphur but four atoms of oxygen. Some salts contain water molecules within the crystals. This is **water of crystallisation**. $MgSO_4.7H_2O$ means that for every one $MgSO_4$ there are seven molecules of water inside the crystal.

6 Where do we write subscripts in formulae?

7 What is the total number of atoms in
a $FeSO_4$;
b $Na_2CO_3.10H_2O$?

Making a salt

BIG IDEAS

You are learning to:
- Understand how neutralisation reactions produce salts
- How to write equations to summarise reactions
- Devise and carry out a salt preparation

Crystals large and small

Crystals of purple amethyst are **gemstones**; we use them to make jewellery. All of the amethyst crystals are the same shape but they may be different sizes. We can make salt crystals using chemical reactions. For each particular chemical even very small crystals can be seen to have the same shape, just use a magnifying glass. The colours of salt crystals vary a lot.

FIGURE 1: Crystals used in jewellery.

1 What is the same when we look at crystals of the same kind of salt?

2 How can you study the shapes of very small crystals?

Neutralisation

There are many examples of neutralisation that fit this pattern.

Few plants can grow well in acid soils. The soil acids can be neutralised by adding a base such as lime. Acid indigestion can be controlled by taking a tablet containing a base.

Base + acid → a salt + water

Salts by neutralisation

The particles inside crystals of a single chemical are all arranged in an identical pattern. This is why the crystal shapes are the same.

We can prepare salts using neutralisation reactions. For example:

zinc oxide + sulphuric acid → zinc sulphate + water

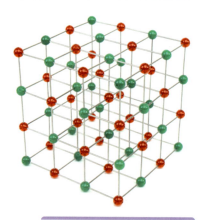

FIGURE 2: Model crystal.

It is important to change all of the sulphuric acid into the new salt. This is why we add an excess of the metal oxide. The excess solid can be removed easily by **filtration** at the end. Evaporation then gives crystals of the new salt, in this case zinc sulphate. If the solution is evaporated quickly using a gas flame, crystals might jump out, we say that it 'spits' crystals. This can be dangerous. Some salts **decompose** when heated strongly, destroying the new salt.

3 Name the salts produced by reaction of sulphuric acid with (a) magnesium oxide; (b) iron oxide.

4 Write a word equation for the formation of copper sulphate from copper oxide.

... decompose

You can make crystals of several salts using neutralisation. Dilute sulphuric acid can be converted into a range of different sulphates.

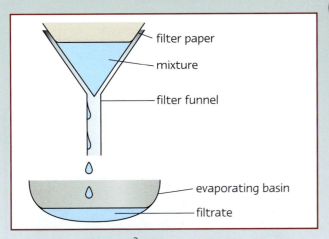

filter paper

mixture

filter funnel

evaporating basin

filtrate

How Science Works

Safety is very important in practical science experiments. Scientists consider the risks before starting a new experiment. Heating a tube of acid in a water bath is safer than heating the tube in a Bunsen flame. The hot gas flame could make the acid boil over and spill. This would create a hazard for all the people working nearby.

HSW

1   Measure 20 cm^3 of acid into a boiling tube.

2   Warm the acid carefully in a water bath at 60–70°C.

3   Add two spatulas of oxide to the tube. Choose from the oxides of zinc, copper and magnesium.

4   Stir carefully. If all the solid reacts, add some more until there is some solid left over.

5   Filter into an evaporating basin.

6   Leave to cool. The best crystals can be obtained by slow evaporation over several days.

7   Record your observations.

Questions

1   Why is the acid warmed before use?

2   Why is it necessary to stir the mixture?

3   Write word equations for reacting zinc oxide, copper oxide and magnesium oxide with the acid.

4   Write symbol equations for reacting zinc oxide, copper oxide and magnesium oxide.

Changing equations

Since neutralisation reactions follow a pattern, we can write equations for many other possible reactions.

Acid	Formula	Salt formed	Example
Nitric	HNO$_3$	Nitrate	NaNO$_3$
Hydrochloric	HCl	Chloride	KCl

5   Write word and symbol equations for the reaction between:
    a nitric acid and sodium hydroxide
    b hydrochloric acid and potassium hydroxide

6   The lime needed to neutralise acid soil is made by heating limestone. Give two environmental problems associated with producing lime.

You don't need to test many metal samples with acids before you notice a pattern of reactions.

Acid + metal $\rightarrow$ salt + hydrogen gas

There are similar patterns when acids react with bases, alkalis or carbonates. These patterns allow us to predict whether particular reactions will work and what the products will be.

A reactive metal

Damaging reactions

Understanding the way that carbonates react with acids is useful for stonemasons and architects. In industrial areas where there is increased acid rain pollution, the use of limestone is a bad idea. Limestone is a carbonate rock and acid particles in the polluted rain react with the rock, destroying it. Since sandstone does not react with acids in the same way, we can predict that sandstones will probably be a better choice of building stone.

1 Which gas is released when acids and carbonates react?

2 How could you identify this gas using a simple test?

3 Name the salts produced when sulphuric acid reacts with (a) sodium hydroxide (b) copper oxide.

4 Predict the products of the reaction between nitric acid and zinc carbonate.

5 Explain why it would be sensible to use an excess of solid zinc carbonate (question 4) and how you could remove it.

6 Predict the shape of the graph that shows how the volume of hydrogen collected changes when magnesium reacts with an acid.

7 In terms of particles, describe what happens in this reaction.

8 Identify any safety hazards in investigating the reaction of metals with acids and explain how you would control them.

ICT Activity

Use the Internet to locate information about these naturally occurring salts (a) calcite (b) barites or heavy spar (c) halite.

Tabulate the information to include crystal shapes and solubilities.

Art Activity

Use coloured discs to represent the ways that particles are rearranged in the reaction of a solid carbonate with an acid. Include a key.

Level Booster

8 You demonstrate a very advanced understanding of particle patterns and behaviour in rocks and in reactions. You can account for anomalies and make confident use of symbol equations.

7 Your answers show an advanced understanding of reaction patterns that you can interpret in terms of particle rearrangements. You can write and interpret symbol equations.

6 Your answers show a detailed understanding of chemical reactions. You can offer explanations based on particle theory and word equations and can make predictions.

5 Your answers show a good understanding of chemical reactions. You can recognise that there are reactions of different types.

4 Your answers show a basic understanding of chemical patterns.

Precipitating salts

BIG IDEAS

You are learning to:
- Explain what precipitation is
- Understand how precipitation can be used to form salts
- Recognise the commercial importance of precipitation HSW

Change places please

Some salts dissolve in water, we say they are **soluble**. Other salts will not dissolve at all, they are **insoluble**. When we mix solutions together, the particles inside can join together in different ways. Imagine that we can mix up people to make new ones.

FIGURE 1: Making a precipitate.

Chris Black + Deepa Blue → Chris Blue + Deepa Black

Lead nitrate + calcium chloride → Lead chloride(s) + calcium nitrate

FIGURE 2: Equation in pictures.

Two new chemicals have been made by mixing. If one of them is insoluble, we see it as a solid powder. This new solid is called a **precipitate**. In equations we show it is a solid by writing (s) after the name, as in the example.

1 What do we call a new solid formed by mixing solutions?

Helping the artists

The colours used by artists are called **pigments**. Some are natural colours and others can be made by precipitation reactions. The first pigment could be made by the following reaction. In the equation, (aq) means aqueous, dissolved in water.

Pigment	Chemical composition	Formula
Barium yellow	Barium chromate	$BaCrO_4$
Chrome yellow	Lead chromate	$PbCrO_4$

The insoluble pigment, barium chromate, can be separated by filtration, washing and drying.

Barium chloride(aq) + potassium chromate(aq) → barium chromate(s) + potassium chloride(aq)

2 Malachite is a green mineral used by artists. What do artists call such coloured materials?

3 Name the products of mixing solutions of calcium chloride and potassium sulphate.

4 Group 1 compounds in the Periodic Table are soluble. Identify the likely precipitate in question 4.

... insoluble ... pigment

You can form lots of insoluble salts by precipitation. Some of them have strong colours and are useful as pigments. Some compounds are toxic, such as those of lead.

By choosing to mix two soluble salts that can form a precipitate, we can make a range of useful new salts. Precipitation reactions are usually very fast. Separating the insoluble salt using filtration leaves the soluble product in the filtrate. This second salt could be recovered by evaporation to give crystals. In precipitation reactions, the two soluble particles from different solutions combine to give a new insoluble product.

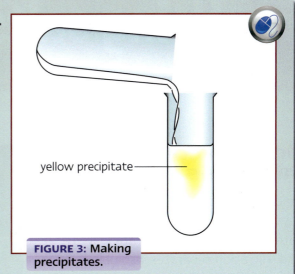

yellow precipitate

FIGURE 3: Making precipitates.

Method:

Solution A	Solution B
a) Lead nitrate	Sodium sulphate
b) Barium chloride	Potassium chromate
c) Calcium chloride	Sodium sulphate

1 Place 2 cm^3 of solution A in a tube and 2 cm^3 of solution B in a separate tube.

2 Mix the solutions and note any changes.

3 Filter the mixture.

4 Open out the filter paper and leave to dry.

5 Describe the appearance of the precipitate.

6 Try different pairs of solutions chosen from the table above and repeat steps 1 to 5.

Questions

1 Name the precipitate formed by mixing solutions in experiment (a).

2 Write a word equation including state symbols (aq or s) for the same reaction.

3 Why should the precipitate be washed with distilled water before drying if the material must be pure?

Using salts

BIG IDEAS

You are learning to:
- Use separation methods based on chemical and physical properties
- Explain the processes that can be used to produce useful materials
- Plan a multi-stage procedure to purify a substance

Separation mixtures

We can separate mixtures in different ways. If we have a solid dissolved in water, just letting it evaporate leaves a solid. Filtration separates insoluble material such as sand. With liquid mixtures, distillation allows us to boil off each liquid separately.

Mixture	How we can separate it
Salt and water	Evaporation
Sand and water	Filtration
Alcohol and water	Distillation

FIGURE 1: The product of separating salt and water by evaporation.

Separation problems

Soluble salts can be separated from a solution by **evaporation**, as with the production of sea-salt. Insoluble salts are separated by **filtration**, followed by rinsing with clean water and drying to give a pure solid.

Ref	Material	solubility
a)	Sodium chloride	Soluble
b)	Calcium carbonate	Insoluble
c)	Potassium sulphate	soluble
d)	Calcium sulphate	insoluble

1 Looking at the table above can you separate the following mixtures? Explain how.
  a Water and compound b)
  b Water and compound c)
  c Water and both compounds b) and c)

... evaporation ... filtration

Purifying salts

Salts found in nature are not pure. Common salt, sodium chloride, occurs as rock salt. This is a mixture of salt, clay and sand or grit. Dissolving the salt in water allows insoluble materials to be filtered off. When the impurities are soluble, such as colours, we need a different method. Mixtures of salts can also be separated if their **solubilities** vary.

Salt formula	Solubility at 40°C as grams/kg of water
KCl	400
NaCl	360
CaSO$_4$	2

A solution containing all three salts would deposit crystals of each separate salt in turn. The least soluble would separate first. This is why natural salt deposits have layers of different salts.

FIGURE 2: Pure salt.

2 Why is filtration alone insufficient to purify coloured rock salt?

3 What would you expect to see if a solution containing the same amounts of salts as in the table were cooled from 50°C to 30°C?

4 In what order would the salts in the table crystallise? Explain your answer.

5 Calcium sulphate has a very low solubility in water and occurs naturally as the mineral gypsum. Some clays contain large gypsum crystals, as on the Isle of Wight. Suggest how these gypsum crystals might have formed.

How Science Works

A sample of rock salt contained sand, sodium chloride and some clay. Plan an experiment to recover both the salt and the sand. Separation techniques are used to give us useful raw materials.

Exam Tip!

Working out similar formulae

Elements in the same vertical column, or group, of the Periodic Table are very similar in their chemistry. If you know the formula of one compound, you can usually work out the others for that group. In group 1 the chlorides include NaCl, KCl and CsCl. In group 2 we find CaCl$_2$ and MgCl$_2$ and in group 3, AlCl$_3$. You don't need to remember every formula, just use the pattern to help work it out. In group 7, called the halogens, the chemistry of the elements bromine and iodine is very similar to that of chlorine. If we know that potassium chloride has the formula KCl, then we can deduce that the other halide salts will have similar formulae. The bromide is KBr and the iodide is KI.

1 Copy the following sentences using the words below. The words may be used once, more than once or not at all.

Bunsen   colours   insoluble   salt   soluble   solute   sun   taste

a   Sea water tastes of _____ .

b   Energy from the _____ evaporates sea water.

c   Blue copper sulphate shows that salts can have _____ .

d   Solutions must contain salts that are _____ .

2 Copy and complete the following table. Write the words **yes** or **no** to show if the named material is a salt.

Material	Is it a salt, yes or no?
Copper oxide	
Sodium chloride	
Sodium hydroxide	
Copper sulphate	

3 Copy and complete the following table, adding the correct state symbols to fit the descriptions. Choose from **s, l, g, aq** or write **none**.

Material	State symbol
Carbon dioxide	
Sodium chloride solution	
Ethanol	
Shaving foam	

4 Copy and complete the following word equations.

a   _____ + base → salt + water

b   zinc + sulphur → ………

c   sodium hydroxide + hydrochloric acid → _____ + _____

d   potassium _____ + sulphuric acid → potassium sulphate
                                                  + water + carbon dioxide

5 Copy the table below and match the two sets of statements by joining each correct pair with a line.

Statement 1		Statement 2	
a	Calcium carbonate can be filtered	**e**	because there are no mobile electrons.
b	Bases may react more slowly than alkalis	**f**	because of solar evaporation.
c	Sulphur fails to conduct well	**g**	because it is insoluble.
d	Saline lakes can produce salts	**h**	because of differing solubilities.

6 Copy the table and identify the relevant particle names, choosing from:

atom   hydrogen particle   hydroxide particle   molecule

Material	Particle present
Alkali	
Carbon dioxide	
Acid	
Argon	

7 Identify the likely precipitate in the following reactions. Choose from:

None   AgCl   BaSO$_4$ CaCO$_3$ KCl   KNO$_3$

a silver nitrate + potassium chloride →

b barium nitrate + potassium sulphate →

c calcium chloride + potassium carbonate →

d copper nitrate + potassium chloride →

8 Solutions of lead nitrate and potassium chromate were mixed to give a yellow precipitate. The mixture was filtered.

a What was the precipitate?

b What was the filtrate?

c Why is distilled water added to precipitates?

d Which of the products is likely to have a colour and why?

9 Copy the following equations and add the correct numbers to balance each equation.

a _____ NaOH + H$_2$SO$_4$ → Na$_2$SO$_4$ + _____ H$_2$O

b H$_2$SO$_4$ + _____ Li → Li$_2$SO$_4$ + H$_2$

c H$_3$PO$_4$ + _____ KOH → K$_3$PO$_4$ + _____ H$_2$O

d K$_2$CO$_3$ + HCl → _____ KCl + _____ H$_2$O + CO$_2$

Learning Checklist

☆ I know how salt can be obtained from sea water.   page 86

☆ I know that metals have special properties.   page 88

☆ I can name examples of metals to show these properties.   page 88

☆ I know that acids and alkalis make salts together.   page 90

☆ I know more than one way to make a salt.   page 102

☆ I know that metallic elements are malleable.   page 88

☆ I can distinguish between bases and alkalis and give examples.   page 90

☆ I know the test for carbon dioxide from carbonate salts.   page 92

☆ I can explain what is meant by precipitation.   page 100

☆ I can explain how the solubility of salts varies with temperature.   page 103

☆ I can write general reactions for the chemistry of acids.   page 93

☆ I can write word equations for salt preparations.   page 96

☆ I can plan an effective method to prepare a range of salts.   page 101

☆ I can predict the identity of precipitates formed by mixing solutions.   page 101

☆ I know that metals usually form basic oxides.   page 91

☆ I know that non-metals usually form acidic oxides and some exceptions.   page 91

☆ I can name salts and interpret their formulae.   page 95

☆ I can summarise salt preparations using both words and symbols.   pages 95-97

☆ I can explain why the reaction rate changes based on the concentration of   page 93
   an acid.

☆ I can write balanced symbol equations for salt preparations.   page 97

Topic Quiz

1 What is left when sea water evaporates?

2 What do we call the opposite of an alkali?

3 Are precipitates examples of soluble chemicals?

4 What is made by reacting acids with alkalis?

5 Name the products of the reaction between potassium hydroxide and sulphuric acid.

6 What would you observe when adding acids to carbonates?

7 How are insoluble salts removed from mixtures with solutions?

8 What is an alternative name for a soluble base?

9 Would you expect sodium sulphate or copper sulphate to be colourless?

10 From which acid could you prepare sodium phosphate?

11 Which properties of metals are associated with mobile electrons?

12 Which metallic property accounts for the formation of wires?

True or False?

If a statement is false then rewrite it so it is correct.

1 Sea water contains sodium chloride.

2 Bases and alkalis are the same thing.

3 Carbonates fizz with acids giving carbon monoxide.

4 The test for carbon monoxide is that lime water goes cloudy.

5 Filtration is useful to separate precipitates.

6 Distillation is a good way to make large crystals from solutions.

7 All metals react with acids to give hydrogen.

8 Alkaline solutions contain hydroxide ions.

9 Most silver salts are insoluble.

10 The salinity of the ocean does not vary.

11 Neutralisation involves the reaction of hydrogen and hydroxide ions.

12 Neutralising all of an acid with a base requires an excess of base.

Literacy activity

Write a paragraph using the word salt in as many different ways as you can. Think about where you come across salts in everyday life.

ICT activity

Research the link between salt mines and the locations of towns and cities. Start by considering the UK, Austria and Poland.

Tanya Streeter

Extraordinary Depths

Diving to the ocean floor on a single breath is an ancient skill used by pearl and sponge fishermen, but only in the past 20 years has it become an internationally competitive sport – free-diving. Free-diving is an extreme aquatic sport, in which divers attempt to reach great depths unassisted by breathing apparatus. It requires superhuman levels of stamina and fitness.

The ultimate no limits world record is the maximum depth reached by a diver on a weighted sled before being pulled to the surface by a lift bag that is inflated by the diver at depth. The world record for no limits free diving is currently held by the American Tanya Streeter. On 17th August 2002 she reached a depth of 160 m. Most people can hold their breath for 40 seconds, Tanya Streeter can hold her breath for over 6 minutes.

Tanya descends on a weighted metal sled at a rate of 3 m/s and by the time she reaches a depth of 50 m her lungs have been squeezed, by the pressure, to the size of a clenched fist. The pressure on her ears causes intense pain which Tanya likens to having two sharp points thrust into her ears. After just over two and a half minutes she reaches the depth of 160 m. Her ascent from the dive is with an air-filled balloon which can produce rapid pressure changes, so she will swim the last few metres unaided as it is at this depth that the pressure changes can be most severe.

At the bottom she loses awareness, nitrogen narcosis has taken a hold. She does not know to release the catch that will return her to the surface. One of the support divers tries to reach her, but is exceeding his safe dive depth limit.

Finally, Tanya releases the catch and rockets back towards the surface. She is in danger of passing out, but cannot be assisted as the judges must see her to be conscious for the record to stand. After an agonising few seconds, she is seen to be fine and the judges confirm the record is good. Six weeks after Tanya Streeter set the world record, her rival Audre Mestre tried to break it. She died by drowning in the attempt as her sled malfunctioned.

Accidents happen in free diving to people who don't practise correctly. Divers can black-out and around 50 divers die each year. The oxygen in the lungs can turn poisonous at depth, making the diver feel drunk, and the crushing water pressure can perforate their eardrums. There are the dangers associated with 'the bends' when divers resurface too quickly. This happens when gases which have dissolved in body liquids and tissues come out of solution as the pressure reduces and form gas bubbles within the body. It is often fatal.

PRESSURE, FORCES AND MOMENTS

BIG IDEAS

By the end of this unit you will be able to explain how forces produce a variety of effects, including pressure and acceleration. You will be able to calculate outcomes such as pressure and moments, and interpret graphs. In your practical work you will have gathered data and repeated readings if appropriate.

What do you know?

1   What is **one** advantage of using a sled to help in the descent?

2   Why do you think the sled is a triangular shape?

3   What is the advantage of using an air-filled balloon to aid in the ascent?

4   What are the dangers of ascending too quickly?

5   What force helps the diver descend so quickly?

6   Why does the pressure on the diver increase as she goes deeper?

7   Why do you think she has large flippers on her feet? What is the disadvantage of wearing them?

8   Why do you think the air-filled balloon has to be inflated at the bottom of the dive?

9   Will the diver continue to accelerate downwards for the duration of the dive? Explain your answer.

Pressure points

BIG IDEAS

You are learning to:
- Understand the relationship between force, area and pressure
- Recognise what pressure is and some examples of high and low pressure
- Understand how ideas about pressure are applied to a range of situations

What is pressure?

A **force** is a push or a pull on an object and is measured in newtons (N). Pressure is caused when a force is applied over an area.

Hitting a nail into a wall puts the point of the nail under high pressure. You have used the hammer to provide a large force. The sharp point of the nail concentrates the force over a very small **area** so the nail can move through the wall.

Some vehicles, like tanks, are fitted with caterpillar tracks. This lets them spread their weight out over a larger area and stops them getting bogged down. Tractors have wide tyres for the same reason, to spread the weight out and lower the **pressure**.

Sharks are very good predators and have been for millions of years. They have very strong jaws and their teeth are very sharp. Because their teeth are so sharp, sharks can concentrate all the force from their jaw into the small area at the tips of their teeth. This high pressure lets them tear flesh.

Snowboarders rely on pressure too. The snowboard is designed not to sink into the snow. The board spreads the weight of the person out over a larger area and this keeps the pressure low. Camels can walk on sand because they have large feet that spread their weight out, lowering the pressure.

1 Why does a nail have a sharp point?

2 Why is it useful for tanks to have caterpillar tracks?

3 Why are sharks able to tear through flesh so easily?

4 Explain how a camel is able to walk on sand even when it is carrying a heavy load.

FIGURE 1: Caterpillar tracks on a tank spread the weight of the vehicle.

FIGURE 2: Sharks' teeth are very sharp.

FIGURE 3: Why doesn't this snowboarder sink into the snow?

... area ... force ...

High and low pressure

If you keep the size of the force the same then:

- the pressure is high when the area is small
- the pressure is low when the area is large.

High pressure	Low pressure
Nail	Tractor tyres
Drill bit	Aeroplane wings
Screw	Building foundations
Ice skates	Snow shoes
Sewing needle	Camel's feet
Bread knife	

5 Explain what pressure means.

6 Give some examples of high and low pressure.

Calculating pressure

Pressure tells us how the force that is applied and the area over which it is applied are related. An equation is used to work out how much the pressure is.

pressure $= \dfrac{\textbf{force}}{\textbf{area}}$ Or in shorthand $p = f/a$

Because force is measured in newtons (N) and the area is measured in square metres the pressure is measured in newtons per square metre (N/m^2). Units of pressure are also called **pascals** (Pa). 1 Pa is exactly the same as 1 N/m^2.

You can use an area measured in cm^2 to calculate pressure but remember that if you do the pressure is calculated in N/cm^2.

Example

What is the pressure on the floor when a man of weight 750 N stands on an area of 250 cm^2?

pressure $=$ force/area

$= 750/250$

$= 3$ N/cm^2

7 **a** Calculate the pressure when a woman of weight 500 N stands in high heels with an area of 1 cm^2 each.

b What area is needed to produce a pressure of 300 Pa from a force of 125 N?

8 Many stately homes insist that women wearing high heels remove them on entry. Explain why.

Pressure in gases

BIG IDEAS

You are learning to:
- Describe what atmospheric pressure is and what causes it
- Understand how particles in gases create pressure
- Recognise how models can be used to explain observations

Pressure all around us

Pressure exists in gases because they are made up of small particles.

The gases in the atmosphere are made of particles that can move. The force of gravity pulls these gas particles downwards towards the Earth's surface. This causes the atmosphere to press down on the Earth and everything on it with a force called **atmospheric pressure**.

The atmosphere is denser nearer sea level than it is at the top of mountains. The higher you go the less gas there is.

The wind is caused by pressure differences in different parts of the atmosphere. **Meteorologists** try to use this information to predict the weather. They map the differences in pressure using **isobars**.

FIGURE 1: Why do mountaineers have to carry oxygen cylinders when they climb high mountains?

FIGURE 2: One of the effects of atmospheric pressure differences.

FIGURE 3: A weather forecaster with a map.

1. Explain why the air exerts a pressure on us. Why do you think it does not seem to affect us?

2. Explain why the atmosphere is more dense at sea level than at the top of mountains.

3. What might limit how high a hot air balloon can get?

... atmospheric pressure ... compressed ... isobar

Particles and pressure

The particles in gases are free to move in all directions. When these particles collide with each other or with other surfaces a force acts on an area causing a pressure.

- The more particles there are in any one space, the higher the pressure is.

 Low pressure

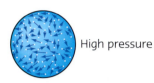 High pressure

- The more collisions there are, the greater the pressure is.
- The more force there is in the collision, the greater the pressure.

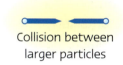

Collision between larger particles

- The smaller the area the collisions affect, the greater the pressure is.

Low pressure – the forces are spread out

Surface

High pressure – the forces are concentrated

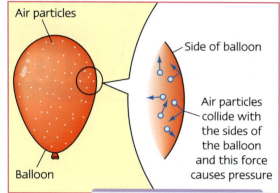

Air particles

Side of balloon

Air particles collide with the sides of the balloon and this force causes pressure

Balloon

FIGURE 4: Pressure in gases.

FIGURE 5: The pressure inside an aerosol is greater than the pressure outside it. When a valve is opened the pressure difference forces the product out.

Because the particles in gases are far apart gases can be squashed or **compressed**. Compressing a gas increases the pressure as the particles now occupy a smaller space. They collide more often and with more force. Gases under pressure are used in aerosols and fizzy drinks.

Fizzy drinks contain carbon dioxide gas. In the can the gas is under pressure and so it is dissolved in the liquid. When the can is opened the pressure is reduced. The gas comes out of solution and expands to create the fizz.

When bicycle tyres are pumped up the pressure rises because there are more air particles inside the tube and they collide with each other and the tyre wall more often and with greater force.

FIGURE 6: Fizzy drinks in a can are under pressure.

4 Explain, using ideas about particles, why gases can be compressed.

5 Explain why the pressure in a bicycle tyre is higher than the pressure in the surrounding air.

FIGURE 7: Pumping more air into a sealed tyre increases the pressure.

Pressure in liquids

BIG IDEAS

You are learning to:
- Understand about pressure in liquids
- Describe how hydraulic systems work
- Explain how hydraulic machines work as force magnifiers

Pressure in liquids

In liquids the particles are packed more closely together than they are in gases with very little space between them. This means that liquids cannot be squashed or compressed.

When we are on land the pressure inside and outside our bodies is the same. For sea divers the pressure outside increases as they go deeper but the pressure inside remains the same. The deeper they are the greater the pressure acting on their bodies.

The pressure increases with depth because the deeper you go the greater the weight of water pushing downwards on you.

In liquids:

- pressure at any one depth is always the same
- the pressure increases as the depth increases
- the pressure acts in all directions.

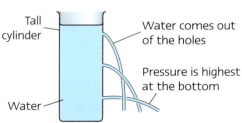

Pressure increases with depth

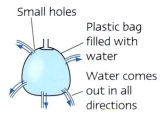

When a plastic bag filled with water is squeezed, the water comes out in all directions

Pressure acts in all directions

FIGURE 1: Pressure in liquids.

1 Explain why pressure increases with depth. Use ideas about particles in your answer.

2 Why do the engineers who build dams need to make sure that the dam wall is thicker at the bottom? Give at least **two** reasons.

What are hydraulics?

Because liquids cannot be **compressed** and the pressure in them acts evenly in all directions, they can be used to transmit forces from one place to another. Systems or machines that are designed to transmit forces through liquids are called **hydraulic** machines.

The basic idea is very simple. The force that is applied at one place is **transmitted** by the fluid to another place.

3 Why can liquids be used to transmit forces from one place to another?

4 What is the basic idea behind hydraulic machines?

How Science Works

Deep-sea submersibles are built to withstand the very high pressures found deep in the seas. Their windows are made of crystal to withstand the pressure. They have enabled people to explore the deep seas.

... compressed ... force-magnifier ... hydraulic

How do hydraulic machines work?

In a simple hydraulic system two cylinders of different sizes are connected together with a tube. Both cylinders and the tube between them are filled with a low-friction liquid, usually oil.

Each of the two cylinders is fitted with a moving **piston** at one of their ends.

Because the pressure is always the same throughout the fluid the pipes do not have to be straight and the pistons can be a long way apart.

It is important that a hydraulic system does not contain air bubbles. If there is an air bubble in the fluid then the force applied to the master piston gets used compressing the air in the bubble rather than moving the slave piston. This makes it much less efficient.

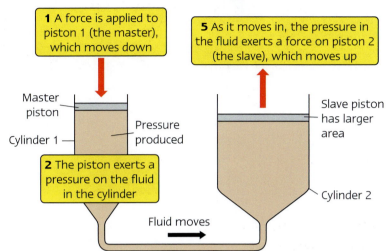

1 A force is applied to piston 1 (the master), which moves down

5 As it moves in, the pressure in the fluid exerts a force on piston 2 (the slave), which moves up

Master piston

Cylinder 1

Pressure produced

Slave piston has larger area

Cylinder 2

2 The piston exerts a pressure on the fluid in the cylinder

Fluid moves

3 This displaces a volume of fluid under the piston

4 The pressure is the same throughout the fluid so the displaced fluid moves through the tube to the other cylinder

5 Why is it an advantage for the connecting pipes to be flexible?

6 Why is it not always an advantage to have flexible pipes? What alternatives could be used?

7 Explain why it is important that a hydraulic system does not contain air bubbles.

FIGURE 2: A hydraulic machine.

Calculating the force produced

We can use the pressure = force/area equation to calculate the forces that can be transmitted by any hydraulic machine.

Worked example

A force of 50 N is applied to a master piston (A) of area 5 cm^2. What force is produced by a slave piston (B) of area 10 cm^2?

The pressure in the fluid = force/area = 50 N/5 cm^2 = 10 N/cm^2

So the pressure at piston B is also 10 N/cm^2 (because liquids cannot be compressed)

The area of piston B is 10 cm^2

The force produced at piston B = pressure x area = 10 N/cm^2 x 10 cm^2 = 100 N

This machine has increased the force produced. It is a **force-magnifier**. But because the larger force at B does not move as far as the smaller force at A, the total amount of energy out equals the total amount of energy put in.

8 a Calculate the pressure produced when a force of 200 N is applied to an area of 20 cm^2.
   b Calculate the force exerted by this pressure on an area of 200 cm^2.

9 Explain why the master piston has a smaller inner surface area than the slave piston.

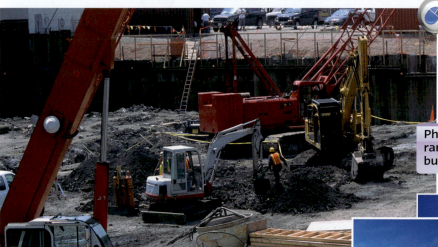

Photo of construction site showing range of large plant machinery and buildings under construction.

Buildings are getting larger and more complex. The time taken to build these ever-larger buildings seems to be decreasing. The main reason for this is the development of new powerful construction machines. The huge machines we see on building sites such as bulldozers, shovels, loaders and cranes all rely on hydraulic systems to work effectively. The development of these machines and the ever-increasing range of new construction materials have enabled rapid advances in building technology.

Hydraulic machines are used whenever a heavy weight has to be lifted or a large object moved. We see them at garages lifting cars to let the mechanics work underneath them. The brakes in cars and even on some mountain bikes work using hydraulics. Hydraulics are useful because they act as force magnifiers. This means that a small input force can be used to produce a large output force. All hydraulic machines rely on the fact that liquids cannot be compressed. So any force applied is transmitted evenly through a fluid. They are different from compressed air or pneumatic machines, such as the air hammers used for road repairs, which rely on compressed air to provide their force.

The largest hydraulic machine in the world is a trencher or rotating shovel which was built by the German company Krupp. It is used in open-cast coal mines. Although at the mine itself the treads are unnecessary, it was cheaper to make the machine self-propelled than to try and move it with conventional hauling equipment.

The machine is 95 meters high and 215 meters long (almost 2.5 football fields in length). It weighs 45 500 tons (that's equivalent to a bumper to bumper line of landrovers 80 miles long) and took 5 years to design and manufacture at a cost of $100 million. Its maximum digging speed is 10 meters per minute and it can move more than 76 000 cubic meters of coal, rock and earth per day.

The world's biggest earth mover.

Assess Yourself

1 When are hydraulic machines used? Think of some examples.

2 What advantages do these machines have?

3 Are there any disadvantages to using them?

4 What does the term 'force magnifier' mean?

5 Explain, using ideas about particles, why pressure can be transmitted through liquids.

6 Explain why most megadiggers have caterpillar tracks instead of wheels.

7 Why do tall cranes not topple over backwards when they are lifting heavy weights?

8 Explain why the atmospheric pressure is higher at sea level than at the top of a mountain.

9 Why is the pressure under water greater than the pressure at the surface?

10 When a construction company wants to build a new skyscraper in a city it needs to consider the design and construction of the structure very carefully. Explain why.

ICT Activity

Use the Internet or the library to undertake some research to find out about the range of hydraulic machines used in the construction industry. For each one say what it is used for and why hydraulics are necessary. You could extend this to give some examples of other places where hydraulic machines are used.

Level Booster

8 You can analyse some of the potential challenges to society associated with the continued increase in construction. You have considered the social, environmental and economic issues associated with the development and introduction of new technologies.

7 You can explain some of the evidence used to support scientific ideas of pressure in liquids and gases and you can give evidence of scientific progress helping to solve social and economic problems.

6 Your answers show a good understanding of the uses of pressure and hydraulics in everyday situations. You can use models to describe the difference between high and low pressure and to describe how hydraulic systems operate. You can describe some of the evidence used to support the idea of pressure acting on objects.

5 Your answers show a good understanding of the uses of hydraulic machines and an awareness of how pressure is used in a range of applications.

4 You can describe some of the uses of hydraulics. Your answers show a basic understanding of pressure and hydraulics.

Turning forces and moments

Spanners

Taking a wheel off a bicycle needs a spanner. The spanner exerts a **turning force** on the axle nut. If the turning force is large enough then it unscrews the nut.

1 Explain why a spanner makes it easier to remove a nut.

Calculating moments

The moment of a force is given by

moment = force applied x distance from the pivot

The moment is measured in newton centimetres (Ncm) or newton metres (Nm).

See-saws turn about a pivot. They move in the direction of the greater turning force. If there is a person on each end of the see-saw they provide opposite turning forces. If the turning forces they supply are equal and opposite then the see-saw will be **balanced**.

Anil wants to move the see-saw in an anticlockwise direction. Mark wants to move the see-saw in a clockwise direction.

The **anticlockwise moment** provided by Anil is equal to

(his) weight x (his) distance from the pivot.

The **clockwise moment** provided by Mark is equal to (his) weight x (his) distance from the pivot.

The see-saw will always move in the direction of the greatest moment.

2 Calculate the direction in which the see-saw will rotate if Mark, who weighs 600 N, sits 2 m from the pivot and Anil, who weighs 500 N, sits 3 m from the pivot.

3 Anil now moves 0.5 m closer to the pivot. Which direction will the see-saw move in now?

How Science Works

Turning forces

If the turning force or **moment** provided by a spanner is not large enough to loosen the nut then there are two ways to increase it.

- Increase the distance from the **pivot** to the force – by using a spanner with a longer handle.
- Increase the amount of force that is used – by pushing harder.

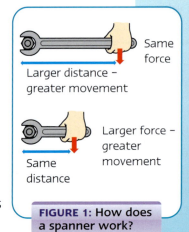

Same force

Larger distance – greater movement

Larger force – greater movement

Same distance

FIGURE 1: How does a spanner work?

So the moment depends on the **force** that is applied and the **distance** from the pivot.

1 What **two** variables do moments depend on?

2 Why do smaller nuts usually only need a shorter handled spanner to undo them?

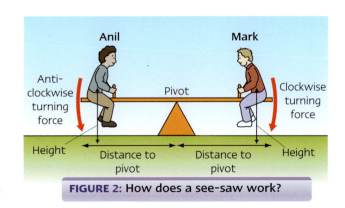

Anil
Mark

Anti-clockwise turning force

Pivot

Clockwise turning force

Height
Distance to pivot
Distance to pivot
Height

FIGURE 2: How does a see-saw work?

... anticlockwise moment ... balanced ... clockwise moment

Principle of moments

If the two moments provided by Anil and Mark are equal and opposite then the see-saw does not move. This is called the **principle of moments**. When this happens the

total clockwise moment = total anticlockwise moment

Example

If Mark now sits 3 m from the pivot, where will Anil need to sit to make the see-saw balance?

When balanced: total clockwise moment = total anticlockwise moment

600 x 3 = 500 x distance. Therefore Distance = 600 x 3 / 500 = 3.6 m

4 What distance must a person weighing 750 N sit from a pivot if a person of 1000 N is sitting 1.5 m from the pivot at the other end in order to balance the see-saw?

5 What weight is needed at a distance of 2 m from the pivot to balance a see-saw with a clockwise moment of 1200 Nm?

FIGURE 3: A very brave man balances his turning forces high above Yosemite Valley in North America.

Did You Know...?

Many of the earliest skyscrapers in New York were built by people who were able to balance on narrow beams without safety equipment. They relied on balancing their turning forces.

FIGURE 4: Tightrope walkers rely on the principle of moments to keep them balanced. Some of them use long poles to help them adjust their balance. By changing the length of pole either side of them they can change the direction of either moment and keep upright.

The Body Machine

BIG IDEAS

You are learning to:

- Explain how organs and tissues in animals function to support movement
- Apply ideas about forces to explain how objects move
- Use appropriate scientific vocabulary

Skeleton and functions – protection, support and movement

The human body is amazingly effective and enables us to do all manner of things with ease. We can use tools, swim, run and move things. Our brain detects a wide range of information and we can respond in a variety of ways.

We can think of our body as a machine. It uses forces, levers and joints, and its structure is well developed.

The **skeleton** is a familiar sight in books about Science. It is the product of millions of years of evolution and has become adapted to carry out a range of functions. In particular it:

- supports the body, enabling us to stand upright.
- allows movement in a range of directions and allows us to apply forces to objects
- provides protection.

1 Look at the diagram of the skeleton. Which parts of it (there are several) enable us to support ourselves?

2 Which parts (there are several) of the skeleton provide protection?

3 Which features of the skeleton make movement easier?

We need to know about forces to understand the skeleton.

4 Draw a simple diagram to show the direction that forces are acting on your lower leg when you are standing upright.

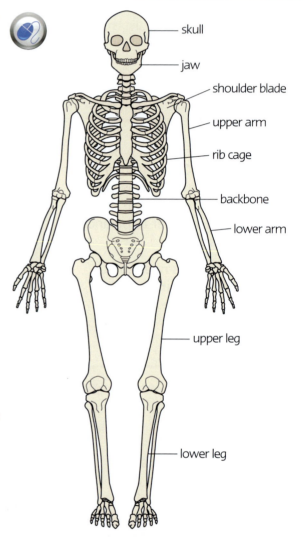

- skull
- jaw
- shoulder blade
- upper arm
- rib cage
- backbone
- lower arm
- upper leg
- lower leg

How Science Works

Scientists often have to work in more than one area of Science. The study of the human body is anatomy and is part of Biology, but the study of forces and movement is part of Physics. To understand about how the skeleton works you need to know about both.

Did You Know...?

The jaw muscle is the strongest muscle in the body. The 1992 Guinness Book of Records states the achievement of a bite strength of 4337N for 2 seconds. What distinguishes the jaw is not anything special about the muscle itself, but its advantage in working against a much shorter lever arm than other muscles.

... antagonistic muscle ... effort ... fulcrum

Muscles working in pairs

There are several joints that have muscles working in pairs. These are called **antagonistic muscles**. Each pulls in the opposite direction to the other. This is because muscles can only contract. This diagram shows the elbow joint with two sets of muscles.

5 Which muscle will make the arm straighten?

6 Which muscle will make it bend more?

7 Why are there two sets of muscles working at this joint?

8 When you nod you are using a joint at the base of the skull and muscles at the back of the neck. Why doesn't this joint need another set of muscles at the front of the neck?

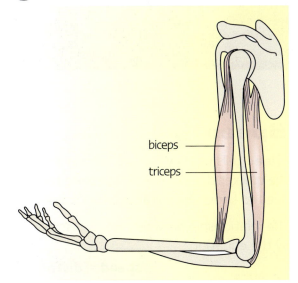

biceps

triceps

Levers and movement

If you use a screwdriver blade to open a can, you are using a **lever**.

These diagrams show the different types of levers. A lever has three features: a **load** – whatever is being lifted or moved, an **effort** – the force that is being applied to move it and a **fulcrum** – the pivot that enables one part to move with respect to another.

Different types of levers have the features in different places.

The body uses levers in several places.

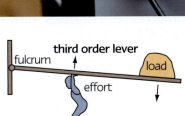

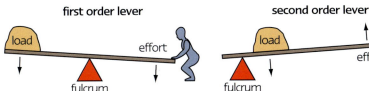

first order lever — load — effort — fulcrum

second order lever — load — effort — fulcrum

third order lever — fulcrum — load — effort

9 Look at the diagram of the elbow joint below right. Where is the load, the effort and the fulcrum?

10 Is this joint like the first, second or third order levers?

11 Identify **three** different joints that have antagonistic muscles and explain which order of lever each of them is.

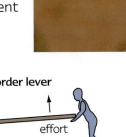

Speedy sums

BIG IDEAS

You are learning to:
- Recognise the relationship between speed, distance and time
- Describe how speed and forces are linked
- Describe what acceleration is

Speed

Speed tells us how fast something is moving. The higher the speed, the quicker the journey. To calculate the speed of a moving object we need to know the distance it has travelled and the time taken.

The relationship between speed, distance and time is given by the equation

$$\text{speed} = \text{distance } (d) / \text{time } (t)$$

We can use the same equation to calculate the distance travelled and the time taken.

distance = speed/time

time taken = distance/speed

We can remember this relationship by using a magic triangle.

Speed is usually measured in metres per second (m/s) or miles per hour (mph). Distances are usually given in metres (m), kilometres (km) or miles. Time is usually measured in seconds.

DISTANCE

SPEED   **TIME**

FIGURE 1: The 'magic triangle'.

1 What does the speed of an object depend on?

How Science Works

A French TGV recently broke the world speed record for a railway train. The TGV was fitted with larger wheels to cover more ground with each rotation, allowing it to reach twice its normal cruising speed.

Linking speed and forces

The speed an object reaches depends on the **forces** that are acting on it. If the forces acting on an object are **balanced forces** then the object will continue to move at a steady speed in the same direction. If the forces are unbalanced then they will affect the object's speed.

Therefore, the speed at which an object travels depends on the size of the forces. It will also depend on the mass of the object.

2 What differences to an object do unbalanced forces make?

Watch Out!

You must always take care to ensure that you are clear which units you are using each time you do any calculations. The units of speed that we use in any answer depend on the units that we have used to measure distance and time. Always give your answer in the correct units.

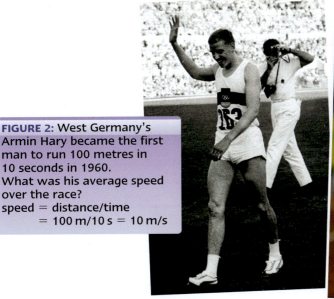

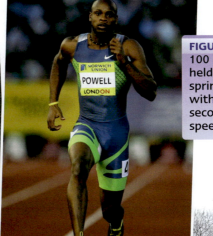

FIGURE 3: The 2007 100 m world record is held by Jamaican sprinter Asafa Powell with a time of 9.77 seconds. What was his speed over the race?

3 How long does it take a lorry to travel a distance of 150 km at a speed of 70 km/h?

4 The Flying Scotsman steam train can travel from London to Edinburgh, a distance of 400 miles, in 4 hours. York is the halfway point in the journey.
  a What is the average speed over the journey?
  b If it arrives in York after 2 hours and 15 minutes, what was the train's average speed from London to York, in mph?
  c What must its average speed be between York and Edinburgh if it is still to arrive on time?

FIGURE 4: The Flying Scotsman.

Accelerating forces

When an unbalanced force acts upon an object in a particular direction the object's speed changes (it **accelerates**) in that direction.

A stationary object will start to move in the direction of the unbalanced force.

An object that is already moving in the same direction as the unbalanced force will speed up. The greater the force the greater the acceleration.

The bigger the mass of the object the greater the force that is needed to provide a particular acceleration.

An object moving in the opposite direction to the unbalanced force will slow down. It **decelerates**. The greater the force the greater the deceleration.

FIGURE 5: Accelerating forces.

5 What is acceleration?

6 What **two** things affect the acceleration of an object?

7 Explain how the acceleration of a moving object can be increased.

8 Explain why an object cannot keep accelerating.

Distance–time graphs

Distance–time graphs

A **graph** is a way of showing how one thing changes against another. It is used to show how two **variables** (the things that change) are related. It provides a type of picture of a journey.

When we go on a journey we can draw a graph to show how the distance from our starting position changes as the time of our journey increases. To do this we need to measure the time taken from the start of our journey and the distance that we have travelled. This **data** then allows us to plot a **distance–time graph** of our journey.

1 What do graphs show us? Why do you think they are useful?

2 What do we need to measure to draw a distance–time graph?

How Science Works

Town planners and road designers often carry out traffic surveys to find out how fast cars are travelling on certain roads. This information helps them to plan new roads.

Graphing a journey

Distance–time graphs show how the time taken to travel any given distance varies during the journey. As soon as you set out on a journey you travel a distance from the starting point. The longer you go the further you travel from the start and the more time that passes. This is shown on Jack's graph.

In part ① Jack walks for 1 mile to his friend's house.

In part ② he waits for 5 minutes for his friend to get ready.

In part ③ they walk half a mile to the bus stop. Because they are talking they do not walk as fast.

In part ④ they are waiting at the bus stop, so they do not travel any more distance for 10 minutes and the line is flat.

In part ⑤ the bus is taking them to school. It takes the bus 10 minutes to travel the 2.5 miles to school. The line on the graph is steeper because the bus is travelling faster.

Distance–time graphs also show us how the **speed** changes throughout the journey.

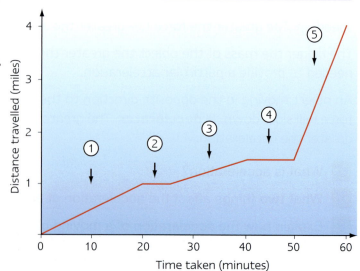

FIGURE 1: Distance–time graph of Jack's journey to school.

... data ... distance–time graph

3 What do distance–time graphs show us?

4 **a** Plot a distance–time graph for the following part of a train journey

Distance (m)	50	150	250	400	600
Time (s)	1	2	3	4	5

b What was the average speed over this time?

c When was the train travelling fastest? How do you know? Work out the speed at this point.

Working out the speed

The journey of a sprint cyclist around a circuit is shown in the distance–time graph below.

The speed of the cyclist changes at different points on the circuit as shown in the graph. The steeper the line is the faster he is travelling. We can work out the speed the cyclist travels on each of the different sections by using the graph.

The average speed of the cyclist's lap is

speed = distance/time = 500/80 = 6.25 m/s

On the long straight his speed = distance/time = 200/20 = 10 m/s

On the tight bends his speed = distance/time = 100/30 = 3.3 m/s

FIGURE 2: Sprint cyclists alter their speed as they go round the track.

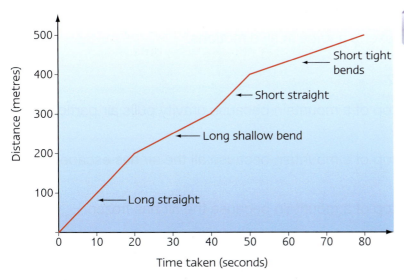

FIGURE 3: Distance-time graph for a sprint cyclist.

5 **a** What was the cyclist's speed through the shallow bends?

b What was his speed on the short straight?

6 Rakib thinks that when the line is steeper the cyclist is accelerating. Rohima thinks that the cyclist is travelling at a constant speed. What do you think? Explain your answer.

Exam Tip!

When you are drawing distance–time graphs always make sure that you have drawn your axes correctly. There should be even divisions on each axis which allow you to plot the data provided. Make sure your graph is big enough to use the space provided on the answer sheet. Make sure you have clearly labelled the axes. Draw the connecting line with a sharp pencil.

1 Describe how you could do an experiment to measure the speed of **three** toys.

a What would you use to measure time?

b What would you use to measure distance?

2 Robbie thinks that an object will only speed up if the forces acting on it are uneven. Rahala disagrees, she thinks that if an object is already moving then an unbalanced force will make no difference to the speed.

Who is right? Explain your answer.

3 Which statement is correct?

a The speed of an object is a measure of its acceleration.

b The speed of an object is the relationship between distance travelled and time taken.

c The speed of an object is a measure of how far it travels.

d The speed of an object is not affected by air resistance.

4 Which statement is incorrect?

a The pressure in a liquid increases with depth.

b The pressure in a liquid decreases with depth.

c The pressure in a liquid exerts a force in all directions.

5 Which statement is correct?

a The air is thinner at the top of a mountain because gravity pulls air particles down.

b The air is thinner at the top of a mountain because all the air has escaped into space.

c The air is thinner at the top of a mountain because the air has frozen.

6 Which statement is correct?

a Acceleration is when an object speeds up.

b Acceleration happens when a balanced force acts on an object.

c Acceleration happens when a balanced force acts on a moving object.

d Acceleration is measured in metres per second.

7 Calculate the speed of the following.

   a A bicycle covering 250 metres in 30 seconds.

   b A skateboard covering 50 metres in 10 seconds.

   c A downhill skier covering 500 metres in 20 seconds.

   d A snail moving 20 centimetres in 3 minutes.

8 Calculate the distance travelled in the following examples.

   a A train travelling at a speed of 20 m/s for one minute.

   b A horse galloping for 30 seconds at a speed of 15 m/s.

   c An arrow leaving a bow at a speed of 75 m/s and taking 1.5 seconds to hit the target.

   d A toy car moving at a speed of 2 m/s for 12 seconds and then a further 15 seconds at a speed of 1.5 m/s.

9 Calculate the time taken:

   a to travel a distance of 100 m at a speed of 5 m/s

   b to travel a distance of 500 m at a speed of 2 m/s

   c to travel a distance of 2 km at a speed of 200 m/s

10 Draw a distance–time graph for the following car journey.

Time (s)	0	30	60	90	120	150	180	210
Distance (m)	0	20	45	65	80	80	120	135

   a What was the average speed over the entire journey?

   b What was the average speed over the first minute?

   c When was the car stationary?

   d When was the car moving fastest?

   e What was its top speed?

11 We can use the principles of moments to calculate turning forces.

   a What does the principle of moments state?

   b Use the principle of moments to decide which way the see-saw will move in each of the pictures.

   c Which see-saw will rotate most quickly? Why?

   b Which see-saw will rotate most slowly? Why?

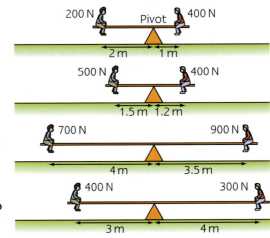

Topic Summary

Learning Checklist

4

☆ I can measure the size of a force and use the right units. — page 110

☆ I can compare the speeds of different things. — page 122

☆ I can describe what speed means scientifically and use the correct units. — page 122

☆ I can use the relationship between speed, distance and time. — page 122

5

☆ I know some key facts about pressure. — page 111

☆ I can describe how to lower pressure by spreading a force out over a larger area and how to increase pressure by concentrating a force. — page 111

☆ I can describe situations where forces are balanced or unbalanced. — page 122

☆ I know that if the forces on an object are balanced, then it moves at a constant speed. — page 122

6

☆ I can describe how pressure works in gases. — page 113

☆ I can explain how hydraulic machines work. — page 115

☆ I can describe how to balance a see-saw. — page 118

☆ I can describe what happens to the turning effect of a force using a longer lever. — page 118

☆ I understand the idea of moments to explain how things balance. — page 118

7

☆ I can calculate pressure using an equation. — page 111

☆ I can use the relationship between pressure, force and area in different situations. — page 111

☆ I can apply the principle of moments to explain situations. — page 119

☆ I know that if an object's speed changes, then the forces acting on it must be unbalanced. — page 123

☆ I know what acceleration is. — page 123

☆ I can draw distance–time graphs and interpret them correctly. — page 125

☆ I can use the equation for speed in calculations and I can convert different units. — page 125

☆ I can explain how turning forces are used in levers. — page 121

128

Topic Quiz

1 What does the term speed mean?

2 What **two** things are needed to work out the speed?

3 How do we measure time? What units are used?

4 How do we measure distance? What units are used?

5 What does the line on a distance–time graph show?

6 Why is the average speed often used to describe a journey?

7 What is meant by acceleration?

8 What happens to the speed of an object if an unbalanced force acts on it?

9 State **three** facts about pressure in liquids.

10 What happens to the moment of a force if the same force is applied nearer the pivot?

11 What is the relationship between force, area and pressure?

True or False?

If a statement is incorrect then rewrite it so it is correct.

1 Speed measures the time taken for something to happen.

2 Unbalanced forces do not affect the speed of an object.

3 The greater the size of the unbalanced force, the slower the acceleration.

4 The units of speed are kilometres.

5 The line on a distance–time graph shows the distance travelled.

6 Acceleration measures how quickly an object changes speed.

7 The air is thinner at the top of a mountain.

8 Ships float because their weight is greater than the upthrust from the water.

9 A see-saw is unbalanced if the clockwise moment is the same as the anti-clockwise moment.

Numeracy Activity

Stopping distances

The distance that a car takes to come to a halt is called the stopping distance. The stopping distance can be split up into two parts – the thinking distance and the braking distance. The thinking distance is the distance that a car travels between the driver deciding to brake and the car actually starting to slow down. The braking distance is the distance that the car travels between applying the brakes and actually stopping. There are a number of factors that may affect both of these. The following table gives some information about stopping distances.

Speed (mph)	Thinking distance (m)	Braking distance (m)	Total stopping distance (m)
20	6		12
30	9	14	
40		24	36
50	15		53
60		55	73
70	21	75	

1 Calculate the missing braking distance figures and thinking distance figures to work out the total stopping distance.

2 Draw a graph to display this data in the most appropriate way.

Did the Ancient Greeks use light as a weapon?

BIG IDEAS

By the end of this unit you will be able to explain how energy is transferred in various contexts, such as light and electrical circuits, and what the effects are. You will have used evidence that you have gathered to construct explanations.

Archimedes (c. 287 BC–c. 212 BC) was an amazing scientist living in Ancient Greece. He was a physicist, an engineer, an astronomer and mathematician. He discovered the principle of buoyancy known as Archimedes' Principle – in his bath!

Archimedes' 'death rays'

When the Romans tried to conquer his home city of Syracuse in 212 BC, Archimedes was put in charge of its defence. Legend has it that he invented giant mirrors that were used to focus light from the Sun on the approaching Roman longboats, causing them to catch fire. These mirrors were sited around the walls of the city. The mirrors were then used to reflect the sunlight back at the Romans' ships which caught fire.

However, no one has managed to find evidence that Archimedes' 'death rays' existed. Scientists, archaeologists and historians have been arguing about it for years. Many have wondered whether the story is a myth. The chances of evidence being found are slim so we will probably never know the truth. Using evidence is important for scientists so that they can be sure about events and their possible explanations.

It is possible that the 'death rays' confused or temporarily blinded the Roman sailors on-board the longboats. Some scientists argue that on a very sunny day the Sun's rays would be enough to set fire to a dry sail made from cloth.

It has been suggested that a large arrangement of highly polished and shaped bronze shields could have been used as mirrors to focus the sunlight on to the longboats and make them catch fire.

Modern scientists have attempted to recreate Archimedes' 'death rays' using models but have had very limited success.

Archimedes was a very skilled scientist.

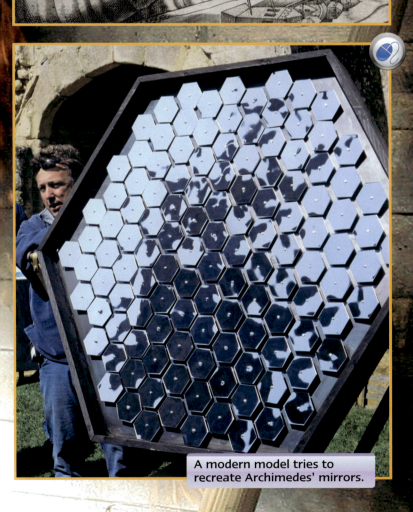

A modern model tries to recreate Archimedes' mirrors.

What do you know?

1 Who was Archimedes?

2 Name **three** things you know about light.

3 Why is evidence needed to prove something has happened?

4 What sort of evidence is useful?

5 Why would the sailors have been blinded by Archimedes' 'death rays'?

6 Would the 'death rays' have worked on a cloudy day?
Explain your answer.

7 What parts of the Roman longboats were more likely to catch fire? Explain your answer.

8 How do shadows form?

9 How do we see an image in a mirror?

10 Why may a catapult have been more useful to Archimedes than the 'death rays'?

Sources of light

Sources of light

We can see the Sun because it gives off (**emits**) **light**. Electric light bulbs and fires are also **sources** of light. Objects that emit light are called **luminous** objects.

Many everyday objects do not give off or emit light – they **reflect** it instead. We see the reflected light from an object when it enters our eyes.

We see the Moon because light from the Sun reflects from its surface and then enters our eyes. We see all the planets in the Solar System because they also reflect light from the Sun.

FIGURE 1: Lighthouses are used to warn ships to stay away from danger. A light source emits light that can be seen from a long way away.

1 Name **three** things that give off light.

2 Why does the shape of the Moon appear to change over one month?

Shadows

The Sun gives out **rays** of light. You can see this by:
- looking at sunlight streaming through a window
- looking at the projector beam at the cinema
- looking at sunlight as it shines through thick clouds in the sky.

A source of light seems to give out thousands of rays in all directions.

FIGURE 2: Why can we see the Moon?

All light travels in **straight lines**. This is why we cannot see around corners. It is also why we get **shadows**. A shadow is formed because light cannot travel around or through an object, unless the object is **transparent**.

The size of a shadow depends on where the Sun is in the sky.
This is why the size of a shadow made by the same object varies at different times of the day.

... emit ... light ... light year ... luminous ... ray ... reflect ... screen

To see a shadow there also needs to be a surface or a **screen**. The size of the shadow that is formed by an object depends on:

- the distance the shadow is away from the light source
- the distance between the shadow and the screen.

Architects and artists think a lot about light and shade in their work.

FIGURE 3: Why does the size of the shadow made by an object change at different times of the day?

3 What causes a shadow?

4 How could the shadow made by an object be made larger?

5 Why does a photographer need to understand about shadows?

How Science Works

X-ray photographs work by producing shadows. X-rays pass through flesh but not bone. This produces a shadow picture of an injury.

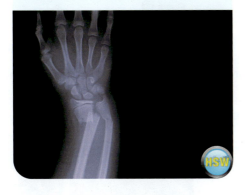

HSW

Light years

Light travels very, very fast. It is the fastest moving thing known in the Universe! It travels at 300 000 000 m/s. It always, or nearly always, travels at the same speed. This helps us calculate how far away very distant objects such as **stars** are. The distances in the Universe are so vast scientists use **light years** to measure them.

A light year is the distance light travels in 1 year.

1 light year is approximately 10 million million km.

- Light from the Moon takes 1.3 seconds to reach us.
- Light from the Sun takes 8 minutes to reach us. This means that we see the Sun as it was 8 minutes ago.
- Light from our nearest star takes 4.3 years to reach us. So when an astronomer looks at it through a telescope they are actually looking back into the past – seeing the star as it was just over 4 years ago!

6 Why do you think light does not always travel at exactly the same speed?

7 What other information would you need to be able to calculate how far away a distant star is from Earth?

Did You Know...?

In calculations the speed of light is taken to be 300 000 000 m/s. But actually the speed of light changes when light passes into different substances. The table below shows the speed of light in some different media.

Medium	Speed (m/s)
vacuum	299 792 000
air	299 700 000
water	225 000 000
normal glass	195 000 000
diamond	125 000 000

... *shadow ... source ... star ... straight line ... transparent ... X-ray*

Light and plane mirrors

BIG IDEAS

You are learning to:
- Describe how light is reflected at plane (flat) surfaces
- Explain some of the uses of mirrors
- Explain how to show the Law of reflection at plane (flat) surfaces

Light and mirrors

When light hits an object it is sometimes **reflected** off the surface of the object.

- Shiny surfaces are better reflectors than dull surfaces.
- The best reflectors are **mirrors**. Mirrors have been used since ancient times when they were usually made from polished metal. As glass became more available it became the main material used to make them.

Mirrors do not absorb light – they reflect it. They can mislead your eyes. This is very useful to magicians and illusionists who use them to make people laugh.

1 a What types of surfaces are the best reflectors?

   b Give **two** examples of good reflectors.

2 Suggest how light is reflected from flat shiny surfaces.

FIGURE 1: A very old mirror and a modern mirror made from glass. Making glass into a mirror involves placing a coating of reflective material behind the glass. Traditionally mirrors were 'silvered', but as silver is expensive, modern, cheaper materials are more widely used.

FIGURE 2: Mirrors can be made that trick your eyes.

Mirror images

You see your **image** in the mirror. The mirror image is:

- the same way up
- the same size
- the same distance behind the mirror that you are in front of it
- back to front.

FIGURE 3: When you look into a full-length mirror what do you see?

When light hits a mirror it is reflected. There are many uses of mirrors other than to let people see how they look. One common use is in a simple **periscope**.

3 What do you think an 'image' is?

4 Is it possible to make an image permanent?

FIGURE 4: Why do some police cars have the word 'Police' written back to front on their bonnets?

... angle of incidence ... endoscope ... fibreoptic ... image

Investigating the Law of reflection

You have already seen examples of mirrors being used in lots of different ways. They also have important uses for safety.

The **Law of reflection** explains what happens when light hits a mirror at different angles. It can be used to understand how mirrors work.

You are going to investigate this law.

Your teacher will provide you with the apparatus that you may need for your investigation.

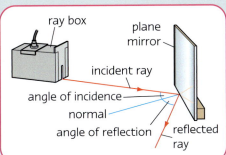

This bicycle helmet has a rear-view mirror built into it. How do you think this increases the rider's safety?

Method:

1 Set up your equipment as shown in the diagram.

2 Measure the **angle of incidence** of your apparatus with a protractor.

3 Shine a ray of light at the mirror so it hits the mirror at the correct angle of incidence.

4 Measure the angle of reflection using the protractor.

5 Repeat **steps 2** to **4** for five different angles of incidence and record your results in a table.

Did You Know...?

Fibre optic lamps are often used to provide lighting effects. When light enters an optical fibre at a certain angle it does not come out again. Instead it continually reflects inside the fibre until it gets to the other end. Fibreoptics are also used by doctors when they want to look inside people without using surgery. The fibre optic instrument they use is called an **endoscope**.

A fibre optic lamp. How does it work?

Questions

1 Describe what you did in your investigation. Include a diagram.

2 What did you find out?

3 What is meant by the 'angle of incidence'?

4 What is meant by the 'angle of reflection'?

5 What is the relationship between the angle of incidence and the angle of reflection?

6 A ray of light hits a mirror with an angle of incidence of 60°. What is the angle of reflection?

7 Why is the incident ray generally brighter than the reflected ray?

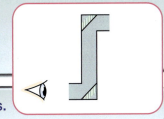

8 Copy and complete the diagram to show how a periscope works.

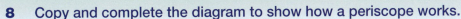

... Law of reflection ... mirror ... reflect ... periscope

Convex and concave mirrors

Not all mirrors are flat mirrors. Curved mirrors are used in a number of different circumstances. There are two types of curved mirrors, convex and concave.

Convex mirrors

A mirror that curves outwards is called a **convex** mirror. You may have seen convex mirrors in shops where they are used to let the shop keeper see the whole shop at one time.

Convex mirrors are also called **diverging** mirrors. They work by reflecting the light rays that hit them so that the rays appear to come from a point behind the mirror. This point is called the principal **focus** of a convex mirror.

FIGURE 1: One use of convex mirrors.

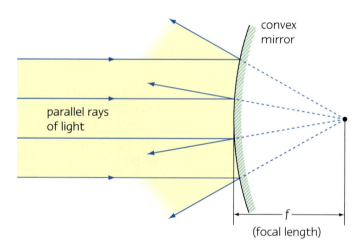

convex mirror

parallel rays of light

f
(focal length)

FIGURE 2: Ray diagram of a convex mirror.

FIGURE 3: Image in a convex mirror.

Convex mirrors always produce an image that is upright and **virtual**. A virtual image is one which cannot be captured on a screen. The image in a convex mirror is also smaller than the object – diminished. This means that convex mirrors give a wider field of view than a plane mirror.

Exam Tip!

When drawing a diagram showing light rays:
- use a ruler and a sharp pencil
- remember to put direction arrows on the rays to show which way the light is travelling
- when light reflects off a surface make sure that you draw the rays just *touching* the surface.

... concave ... converging ... convex

Concave mirrors

A mirror that curves inwards is called a **concave** mirror. Concave mirrors are often used to magnify an image.

They work by collecting parallel light rays and reflecting them through a principal focus, F. Because they work to bring light rays to a focus they are also called **converging** mirrors.

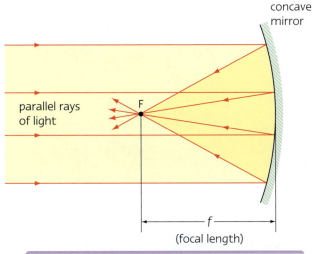

FIGURE 5: Ray diagram of a concave mirror.

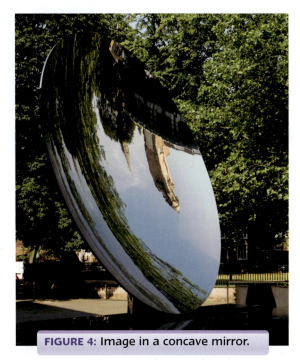

FIGURE 4: Image in a concave mirror.

1 Why are convex and not concave mirrors used in shops and on buses?

2 What is meant by the term 'principal focus' in a convex mirror?

Focal length of concave mirrors

The distance from the focus to the mirror is called the **focal length**. The image in a concave mirror depends on where the object is in relation to the focal length. If the object is closer to the mirror than the focal length then the image is upright and magnified. A mirror with a shorter focal length is more powerful because it bends the light more. This is the principle of shaving mirrors and make up mirrors.

FIGURE 6: A shaving mirror.

3 What happens to the focal length of a concave mirror as the curvature of the mirror is increased?

4 What effect will this have on the image produced?

5 Why are concave mirrors used in solar reflectors?

6 Explain why the image in a concave mirror is upright and magnified, when the object is placed closer to the mirror, than the focal length.

Did You Know...?

Concave mirrors are also used to collect sound, heat and radio wave energy and focus it onto a point. This is the principle used in satellite television receiver dishes and radio telescopes.

In the opposite way, energy can be sent outwards from concave mirrors. This principle is used in electric fires, and bicycle and car lights.

... diverging ... focus ... virtual ... focal length

Total internal reflection

BIG IDEAS

You are learning to:
- Describe and explain total internal reflection
- Explain how a fibre optic works
- Describe some of the uses of fibre optics

Fibre optics

You may have seen lights like these. They use glass fibres to transmit light energy from one place to another.

Fibre optics are not just used in domestic lighting; they are a vital component of the modern communication systems that are used across the world every day. These fibre optics use light signals to carry data, voices and images. Fibre optics have virtually replaced traditional copper wire in telephone lines and they are increasingly used to 'hard wire' computers over local networks.

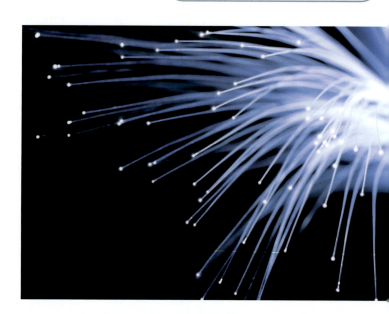

1 What are fibre optics used for?

2 Why are they increasingly used instead of traditional copper wire in telephone cables and computer networks?

Total internal reflection

The way that light behaves when it enters glass depends on the shape of the glass and the angle at which the light enters the glass. The light may be refracted at both glass surfaces and emerge, or it may be reflected in whole or in part from the second surface.

incident ray

Totally internally reflected ray

Angle of incidence (7) greater than the critical angle

If the **angle of incidence** is increased so that it is greater than the **critical angle** (42° for glass), then the light ray is not refracted at all at the second surface. Instead it is totally internally reflected and emerges on the same side that it entered.

Total internal reflection only happens when the ray of light is travelling from a denser material such as glass to a less dense material such as air.

3 Explain what is meant by total internal reflection.

FIGURE 1: Diamonds are cut to make use of total internal reflection (TIR).

... angle of incidence ... critical angle

4 Copy and complete the **two** diagrams to show what happens when light enters a prism.

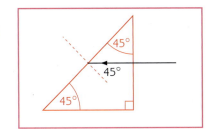

How do fibre optics work?

A fibre optic is a strand of glass, with dimensions similar to those of a human hair, surrounded by a transparent cladding. They are arranged in bundles called **optical cables** and used to transmit light signals over long distances.

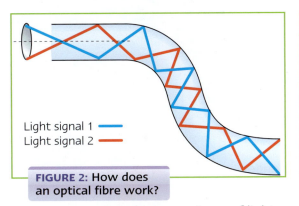

Light signal 1 ——
Light signal 2 ——

FIGURE 2: How does an optical fibre work?

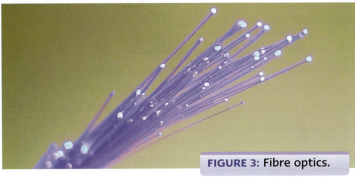

FIGURE 3: Fibre optics.

A ray of light can be passed down a glass fibre providing that the light ray always strikes the internal surface of the fibre optic at a certain angle. This prevents the light from escaping out from the sides of the cable. Instead of escaping, the light is totally internally reflected through it. This means that the light energy that enters the cable at one end is transmitted through the whole length of the cable before emerging at the other end. By ensuring that different beams of light enter the fibre optic at slightly different places, lots of separate signals can be transmitted down one individual fibre. This is an example of the way in which light can be used to transfer energy from one place to another.

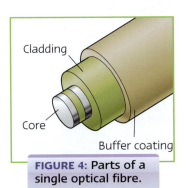

Cladding

Core

Buffer coating

FIGURE 4: Parts of a single optical fibre.

Hundreds or thousands of these optical fibres are arranged in bundles in optical cables. Since they are lightweight and flexible, this allows more strands to go through the same cable, so more information can be carried. There is less signal interference, they are non-flammable and they require less power to transmit signals through them, so they are ideally suited for carrying digital signals.

Fibre optics are also used in many flexible digital cameras since they are so flexible and can transmit and receive light.

5 Explain how a fibre optic transmits information along great distances.

6 Research some of the other advantages of fibre optics.

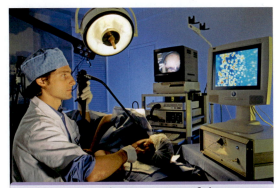

FIGURE 5: An endoscope – one of the many uses of fibre optics. Can you think of others?

... optical cables

Refraction

BIG IDEAS

You are learning to:
- Describe how light is refracted at plane (flat) surfaces
- Recognise examples of refraction

Bending of light

Light travels much faster than anything else known. It travels at slightly different speeds in different materials (see page 133). So when light travels through an object its speed changes.

When light travels from air into a **denser**, **transparent** material such as water, glass or diamond it slows down. When it leaves the denser material it speeds up again. This slowing down or speeding up may cause the light ray to bend. This bending is called **refraction**.

1 What happens to the speed of light when it enters a denser material?

2 What does 'refraction' mean?

Explaining refraction

When a light ray passes from air into a glass block it slows down and it may be refracted. The amount of refraction that occurs depends on the angle at which the light ray hits the glass block.

3 Explain why a ray of light travelling along the normal does not refract when it enters or leaves a transparent material.

4 When light travels at an angle into a denser material why does it bend towards the normal?

A Light entering the glass at an angle of 90°

Air
Glass

The light ray slows down when it enters the more dense glass. It enters the glass along the **normal**. Its direction does not change

B normal

Air
Glass

Light enters the glass block at an angle, it slows down in the glass and refracts towards the normal

C

Glass
Air

As the light ray leaves the glass block, it speeds up and refracts away from the normal

FIGURE 2: A closer look at refraction.

FIGURE 1: Refraction in action.

Did You Know...?

Refraction is also used by diamond cutters in the jewellery trade to make a diamond appear more impressive.

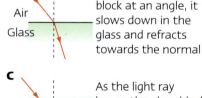

... concave ... convex ... dense ... focus

Examples of refraction

Refraction of light can be very useful. One of its common uses is in **lenses**. A lens is used to **focus** light. There are two types of lenses – **convex** and **concave** – used for different purposes.

- Convex lens – focuses light rays at a point.
- Concave lens – spreads out light rays.

The amount of refraction that happens in a lens depends on the angle at which light rays hit it. Examples of uses of lenses are in spectacles and contact lenses that help improve people's sight.

5 What is the difference between a convex lens and a concave lens?

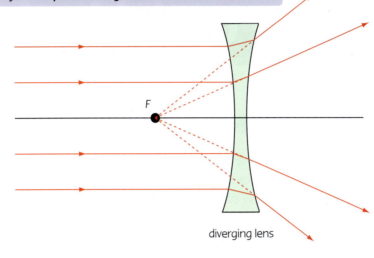

FIGURE 3: What happens to light rays that pass through a convex lens and what happens to rays that pass through a concave lens?

F

diverging lens

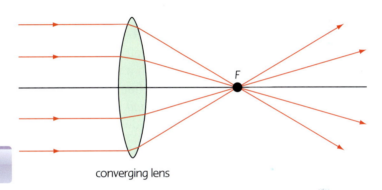

F

converging lens

FIGURE 4: Refraction through spectacles. The light is refracted and focused by the lens.

How Science Works

'Cats eyes' is the name of the objects in the middle of roads that help drivers see at night. They contain small glass beads protected by a flexible rubber casing. The front surface of each bead is clear and works as a lens that focuses incoming light rays on to a rear mirrored surface, which then reflects the light back at the driver.

1 Do Cats eyes in roads rely on refraction, reflection or both? Explain your answer.

2 Cats eyes are usually white at night. Sometimes they can be red or green or even blue.
   a Can you think how they are coloured?
   b Can you give **two** reasons why road designers may use differently coloured Cats eyes?

Exam Tip!

When you draw a refraction diagram remember to:
- show the light ray refracting towards the normal as it passes from the air into glass
- show the ray refracting away from the normal as it leaves the glass
- make sure all your lines connect up and have direction arrows
- draw the normal line (a line at right angles to the glass surface).

... lens ... normal ... refraction ... transparent

Isaac Newton 'the optickian'

BIG IDEAS

You are learning to:
- Describe how a prism affects white light
- Recognise that the amount of light refracted depends on its colour
- Describe how scientific ideas develop from experimental evidence

HSW

Sir Isaac Newton

Sir Isaac Newton is one of the most famous scientists ever. He was able to explain how things work by using the results from his experiments. You may have heard that an apple fell on his head and this helped him describe gravity and that he invented the reflecting **telescope**. He explained about movement and forces. He was also an alchemist – someone who tries to find a way to turn lead into gold. He also spent some of his early career investigating light.

One of Newton's experiments involved shining a beam of light into a prism. Newton used sunlight for his light source (electricity was yet to be discovered). He made a beam of sunlight by cutting a slit in the shutters in his room. He then aimed this beam of light at a large glass prism that had a screen behind it.

> ### Exam Tip!
> When you draw rays of light through a prism remember that they refract when they enter the prism *and* when they leave the prism. Always use a ruler to make sure you draw the rays of light as straight lines.

Newton noticed that sunlight (or **white light**) was split up by the glass of the prism into many different colours of light. These different colours are the colours of light seen in a rainbow and they are called a **spectrum**. This splitting of white light into its component colours is called **dispersion**.

Newton tried hard to see if he could split these colours of the spectrum any further but he could not. He did succeed by using a lens and a prism to recombine the colours to make white light once again.

FIGURE 1: Newton splitting white light into its spectrum of colours. Can you say the order of colours without looking? (Hint: use the mnemonic 'ROYGBIV' to help you.)

Did You Know...?

Sir Isaac Newton (1643–1727).

Early in his career Newton experimented with light and discovered some very interesting facts. They were largely ignored at the time until in 1704 when Newton wrote his book *The Opticks*. This is where we get the word **'optics'** from – the study of light.

Close to the end of his life Newton wrote in his memoirs: "I do not know what I may appear to the world, but to myself I seem to have been only like a boy playing on the sea-shore, and diverting myself in now and then finding a smoother pebble or a prettier shell than ordinary, whilst the great ocean of truth lay all undiscovered before me." In 2005 The Royal Society voted Sir Isaac Newton to have been more important in the history of science even than Albert Einstein – Newton was a truly amazing man.

... dispersion ... Electromagnetic Spectrum ... optics ... refract

He concluded that white light must be a mixture of different colours.

1 Give **two** reasons why Newton had to cut a slit in the blinds of his room.

2 What did Newton observe when he did his light and glass prism experiment?

3 a What are the colours of the spectrum?
  b How can you remember the order of the colours?

How does a prism work?

When a beam of white light hits a prism it **refracts**. Because the colours in white light have different **wavelengths** they refract by slightly different amounts. The spectrum seen is a result of the different amounts of refraction of the different wavelengths in light.

4 Why do the different colours of light refract by different amounts when they hit a prism?

5 Why does a spectrum appear?

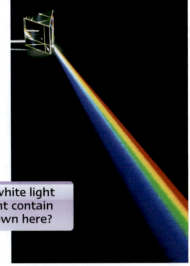

FIGURE 2: Explaining the white light spectrum. Does white light contain any other colours not shown here?

Explaining dispersion

All the colours in white light travel at the same speed in a **vacuum**.

When the different wavelengths of white light enter transparent glass they all slow down and refract by different amounts. How much the light refracts depends on its colour.

Red light has a longer wavelength so it slows down less and is refracted through a smaller angle than violet light which has a shorter wavelength and is refracted more. Since red light refracts less than violet light, they are separated or **dispersed** by the prism and emerge as individual colours.

6 Explain what 'refraction' means.

7 Explain, using a diagram, how white light is dispersed by a prism.

Electromagnetic waves

Light waves are part of a larger family of waves called the **Electromagnetic Spectrum**. All the members of this family, which include X-rays, microwaves and radio waves, transmit energy from one place to another.

8 Use the Internet or the library to find out about the Electromagnetic Spectrum. What properties do all electromagnetic waves have in common? For each type of wave, find a use.

Colour

BIG IDEAS

You are learning to:
- Understand that filters and coloured objects absorb some colours (and transmit or reflect others)
- Describe what effect coloured filters have on coloured objects

Coloured objects in white light

Different objects **reflect** different **colours** of light. The colours that are not reflected by an object are **absorbed** (taken in) by it.

FIGURE 1: Colour is all around!

> The colour of an object that we see is the colour of the light that it reflects.

1 What colours of light are reflected in the scenes in Figure 1?

2 What do you think may happen to the rest of the colours that are not seen in Figure 1?

3 Which colours of light are absorbed by a red bus?

4 Explain why we see the colour of grass as green.

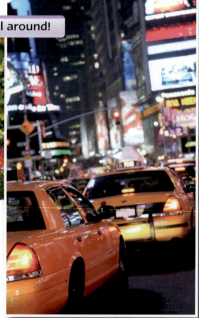

Black and white

Most objects reflect more than one colour. The colours that an object reflects depend on the object.

- A purple object reflects purple light. But purple is made from red and blue so it also reflects red and blue light.

- A white object reflects all the colours and absorbs none.

- A black object is the opposite to a white object. It absorbs all the colours and reflects none. So you could say that black is the absence of colour or the absence of reflected light.

5 How do we know that most objects reflect more than one colour of light?

6 What colour would a red book be in:
  a red light
  b white light
  c green light
  d blue light?

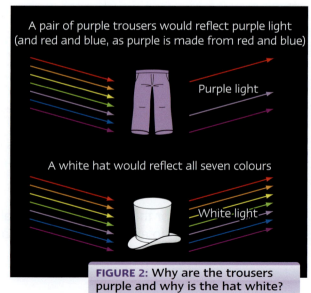

A pair of purple trousers would reflect purple light (and red and blue, as purple is made from red and blue)

Purple light

A white hat would reflect all seven colours

White light

FIGURE 2: Why are the trousers purple and why is the hat white?

... absorb ... colour ... coloured filter

Coloured filters

A **coloured filter** allows some colours of light to pass through it and absorbs the rest. We see the colour of light that the filter allows through or **transmits**. Coloured filters transmit light that is the same colour as the filter.

- A red filter transmits only red light and absorbs all the other colours (orange, yellow, green, blue, indigo and violet). We see red.
- A blue filter transmits only blue light and absorbs all the other colours (red, orange, yellow, green, indigo and violet). We see blue.

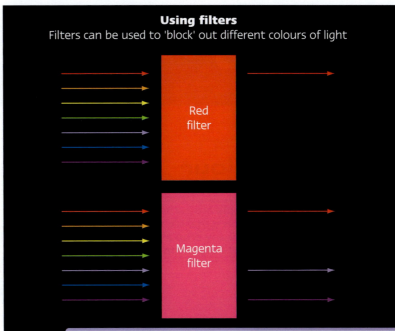

Using filters
Filters can be used to 'block' out different colours of light

Red filter

Magenta filter

FIGURE 3: What would you see through a green filter?

Combining filters

If two coloured filters are put together the light can be totally absorbed. For example, when a green filter is placed in front of a white object it only lets green light through and absorbs the rest. If a red filter is then placed in front of the green one, the green light does not pass through the red filter and the object appears black.

7 Explain why a blue object appears black when viewed through a red filter.

8 Explain why a red object appears red when viewed through a red filter.

Did You Know...?

Imperial Purple was a natural dye used by the Romans. The dye came from a shellfish called murex. The dye was very expensive because of the large number of shells required to make it and the hard work it took to prepare the dye. More than 1200 shells were used to make 1 g of the purple dye.

How Science Works

The food industry uses the connection between taste and vision. Food colourants are added to ensure that the colour of the food matches our expectations, for example, butter is naturally much whiter in colour (similar to the colour of lard) and the canning process deprives the garden pea of much of its natural colouring.

HSW

Using colour

BIG IDEAS

You are learning to:
- Explain the difference between the primary and secondary colours of light
- Describe some uses of colour in everyday life
- Explain why coloured objects look different in differently coloured lights

Everyday colour

When people are in the spotlight on a theatre stage they are clearly seen as the light reflects only from them.

Bright **colours** are used as **warnings** by humans and in nature to warn other animals to stay away.

Camouflage is used by soldiers so they blend into the background. Most camouflage works by breaking up the solid edges of objects so they are much harder for an observer to see. In nature some plants and animals survive by using camouflage to make them difficult to see. They use a range of colours to let them do this. The chameleon can actually change colour to blend into its surroundings!

1 Give **three** examples of bright warning signs used by people.

2 What does 'camouflage' mean?

3 Give **three** examples of where camouflage is useful.

The primary colours of light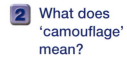

FIGURE 1: When all three primary colours of light are mixed, what colour is made?

At the beginning of the 19th Century, Thomas Young (1773–1829) investigated the effect of mixing coloured lights and found that almost all colours, including white, could be produced using different combinations of just three colours – **red**, **green** and **blue**. He called red, green and blue the **primary** colours of light. White is made by mixing the three primary colours of light.

4 What are the primary colours of light?

5 Why do you think they are called the primary colours?

Adding colours

White light can be split up to make separate colours. These colours can be added together again
The primary colours of light are red, blue and green:

Adding blue and red makes magenta (purple)

Adding blue and green makes cyan (light blue)

Adding red and green makes yellow

Adding all three makes white again

Watch Out!

The primary colours of light are red, green and blue. This is not the same as paints that have different primary colours.

Mixing the primary colours of light

White light can be split up to make separate colours. These colours can then be combined together again to make a range of different colours.

When the primary colours of light are mixed together a different set of colours is made.

Cyan, **yellow** and **magenta** can be made by mixing together two primary colours of light. They are called the **secondary** colours of light. All the colours of light are seen in the same way by **reflection**.

6 Copy and complete the table below.

Mix of primary colour of light	Secondary colour of light produced
red + blue	
red + green	
green + blue	
blue + yellow	
red + green + blue	

7 Explain the difference between reflection, refraction and dispersion.

8 Suggest why in the theatre set designers need to talk to lighting engineers.

The effect of coloured light

When coloured objects are looked at in differently coloured light they appear to be a different colour.

- In red light a red book reflects red and appears red.
- In red light a blue book appears black because there is no blue light for it to reflect.
- In red or blue light a green book appears black because there is no green light for it to reflect.

9 Explain why a red book appears black in green light.

FIGURE 2: Designing the set for a rock concert. The lighting engineers need to work closely with the music director to achieve dramatic effects.

Did You Know...?

Coloured objects can also be seen on a television screen or VDU by light emitted from red, green and blue phosphors. A white colour appears when all the phosphors emit equal amounts of red, green and blue. This is called additive colour mixing.

A plasma TV screen showing the different phosphors.

Seeing the light

BIG IDEAS

You are learning to:
- Recognise that light travels from light sources to our eyes so we can see them
- Explain how the eye works
- Describe how eyes have evolved

The evolving eye

We only see things when light **reflects** from an object and enters our **eyes**. We now know that our eyes are only **sensitive** to certain types of light. For example, we cannot see ultraviolet light whereas bees can.

Eyes have existed for a very long time. One of the earliest eyes belonged to an ancient creature called a trilobite, a sort of giant arthropod. Its eye was made from a mineral called calcite and it was in a fixed position. It had hundreds of different lenses. Eyes with more than one **lens** are called **compound** eyes.

Compound eyes are still common in nature. Nearly all **insects** have compound eyes.

Humans and many animals have eyes with a single lens.

How Science Works

Charles Darwin, who explained evolution, considered the eye such a perfect tool that he doubted whether it could ever just have evolved by chance. It caused him such problems with his world-changing evolution theory that he nearly didn't publish his findings. It's a good job he did. He gave us what is arguably the greatest scientific theory – that of evolution – ever.

Charles Darwin (1809–1882).

Did You Know...?

The kestrel is able to see ultra-violet light. Its favourite food is voles that are often found in roadside verges, well camouflaged in the long grass. The voles mark their territory with their urine. Their urine emits ultraviolet light that a kestrel is able to see. It then watches for movement, allowing it to catch the vole!

This kestrel is breakfasting on vole.

FIGURE 1: a A fossil of the ancient trilobite that lived about 500 million years ago and **b** a horsefly's eye. Both organisms have compound eyes. How are compound eyes different to human eyes?

1 Suggest what problems there may be with having compound eyes.

Complex eyes

Eyes all work in the same basic way. Over time they have evolved into lots of different shapes and sizes.

2 Why do you think there are so many different types of eyes in animals?

3 What is the purpose of the lens in a complex eye?

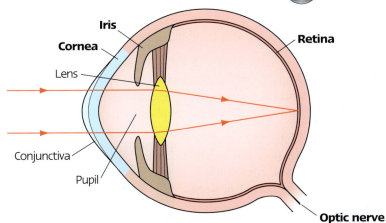

FIGURE 2: All these animals have **complex** eyes. In a complex eye there is one lens that focuses light entering the eye on to a light-sensitive retina at the back of the eye.

How the eye works

FIGURE 3: How the complex eye works.

Labels: Iris, Cornea, Lens, Conjunctiva, Pupil, Retina, Optic nerve

Did You Know...?

The retina contains two types of microscopic light-detecting cells called rods and cones.

- Rods see well in dim light but do not see colours.
- Cones work only in brighter light and see colours and fine details.

All these cells are in the curved sheet of the retina (see Figure 3) that is about as big as the tip of your thumb.

4 Draw a flowchart to explain how a human eye works.

5 **a** Why is it important to control the amount of light entering the eyes?
b What could you do in a bright place to protect your eyes?

When something goes wrong with our eyes we have a sight examination with an optician.

They use a chart like this. The letters get smaller lower down the chart. If people cannot read it all then glass lenses are used to let them see even the smallest letters.

Many people wear **glasses** or contact lenses. Glasses have been used for many years but contact lenses are much more recent. Early glasses were less effective than modern ones because it was not possible to grind the glass to exactly the right shape.

Modern materials allow much greater accuracy. Glasses work by refraction: by getting the right amount of refraction they can work with the lens in the eye to focus the light clearly onto the retina.

Artificial lenses are made from glass or plastic. They help the lens in the eye focus the light correctly. People who wear glasses usually have either short sight or long sight. By using lenses of the right type and strength, the problem can be solved. Short sight is cured with a concave lens. Long sight is cured with a convex lens. Without their glasses, the chart might look like this.

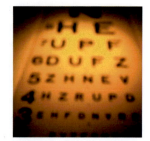

Not all eye problems are as easily solved. Sometimes glasses do not work and surgery is needed. Commonly eye surgery is used to treat:

- cataracts – when the lens becomes cloudy
- glaucoma – where the pressure in the eye builds up
- detached retinas

A cataract is an abnormal cloudy and opaque area that can form in the lens of an eye. The condition will lead to blindness unless it is treated. A cataract operation involves cutting the cornea (at the front of the eye), removing the cloudy lens, and replacing it with a plastic fixed-focus lens. The lens is carefully machined to provide just the right amount of refraction.

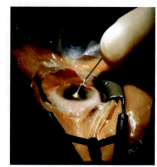

A cataract operation.

This is an increasingly available option these days and many more people are choosing to have this treatment. Doctors can use a laser to permanently change the shape of the cornea. The cornea is the fat clear front of the eye which covers the coloured iris. It is responsible for some focusing of the light which enters the eye. The lens, which is just behind the pupil of the eye, focuses the rest of the light. The laser is directed on the cornea, the outer, clear, round structure that covers the coloured part of the

Laser eye treatment.

eye (iris) and the pupil. The laser uses light energy to change the shape of the cornea and so bring the image into focus. For people with short-sightedness, for instance, the laser is set to reduce the thickness of the cornea. This enables the eye to bring images into focus properly.

Lasers can also be used in the treatment of detached retinas.

More recently eye surgeons have begun to transplant corneas and replace lenses and corneas with artificial ones. These are still complicated treatments.

Until very recently the only way to get a new cornea was by a transplant from another person. It is now possible to grow corneas in laboratories. This has increased the availability of the operation.

A cornea growing in a laboratory.

Assess Yourself

1 What is an optician?

2 Why do some people have difficulty seeing properly?

3 What is the difference between long and short sight?

4 Why have contact lenses only been available recently?

5 Name **three** eye problems that glasses cannot cure.

6 What is a cataract?

7 Explain why people with cataracts may not see clearly.

8 Laser eye treatment is becoming common.

   a What are the main reasons why people are choosing to have this treatment?

   b Can you think of any dangers with it?

9 Artificial lenses are still in their infancy. What difficulties would need to be overcome for their use to become more widespread?

10 What advantages are there in growing corneas in laboratories rather than using transplants?

Art activity

Over many years artists have tried to capture the appearance of real objects and real places. Choose two or three pieces of work done by artists and consider how they have used different colours to produce their work. How have they used shadows and light to enhance the appearance of the things they are painting?

Media Studies

How does the media use colour in the way it presents news reports or advertising campaigns? What impact are these colours designed to have? How could you design a survey to find out how effective the use of colour in advertising campaigns or newspapers is?

Level Booster

8 You recognise how sight correction techniques have evolved over time as technological advances have led to a greater range of available options and you can explain why different treatments are used in different circumstances.

7 Your answers show an advanced understanding of advanced sight correction, a grasp of scientific terminology, an advanced appreciation of how vision difficulties occur and how they are resolved and the effect this has. Your answers also give evidence of scientific progress helping to solve social problems.

6 Your answers show a good understanding of sight correction, a good grasp of scientific terminology, and a good appreciation of how the range of surgical and non-surgical corrective procedures work and their effect on people's lives.

5 Your answers show a good understanding of sight correction, a good grasp of some scientific terminology, and some appreciation of how vision problems are resolved.

4 Your answers show a basic understanding of sight correction and a basic grasp of some scientific terminology.

Resistance in circuits

BIG IDEAS

You are learning to:
- Explain what electrical resistance is
- Describe the factors affecting resistance in circuits
- Recognise that the resistance of wires varies with their length

Electrical resistance

We can use any type of metal wire to connect up components in an electrical circuit because all metals conduct electricity. But the amount of **current** that flows in a circuit is affected by the type of metal.

All the components in a circuit transfer electical energy into different forms. A component **resists** (opposes) the movement of a current through it. **Resistance** is different in every material. Insulators have a very high resistance which is why a current cannot pass through them and energy is not transferred through them.

The lower the resistance of a wire, the higher the current flowing through it will be. Conducting materials such as copper and aluminium have a low resistance and so current passes through them easily. They are used to carry large currents.

1 Give **two** examples of metals that have a low resistance.

2 Try to describe what you think 'resistance' means.

What affects resistance?

The resistance of a wire depends on:
- what material it is made from
- how long it is (short wires have less resistance)
- how thick it is (thick wires have less resistance).

A short thick copper wire has a very low resistance. When a current travels around a circuit it is important that as little **energy** as possible is used to heat up the connecting wires. This is why copper wire is used in most electrical cables.

3 Why is high-resistance wire used for the heating element in a kettle?

4 What does the resistance of a wire depend on?

5 Suggest why aluminium is preferred to copper in overhead power lines.

How Science Works

Because long wires have a higher resistance than shorter wires it is important that when electricity is transported around the country in the National Grid, low-resistance wires are used. This keeps heat loss in the wires to a minimum. But the birds can still warm their toes!

Did You Know...?

Thomas Edison didn't invent the first light bulb – but he did invent one that stayed lit for more than a few seconds.

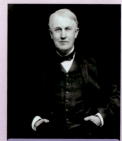

Thomas Edison, 1847–1931.

Thomas Edison invented more than 2000 new products, including almost everything needed for us to use electricity in our homes: switches, fuses, sockets and meters.

FIGURE 1: The high resistance in the filament wire makes the light bulb glow white-hot.

.. current ... energy ... nichrome

Investigating resistance

When a current is pushed through a high-resistance wire, such as **nichrome**, a lot of the energy supplied by the battery is transferred into heat in the wire. High-resistance wires transfer more energy into heat. So, they are used in the heating elements of kettles, electric fires, hair dryers and in some light bulbs.

You are going to investigate how the length of nichrome wire affects its resistance.

Your teacher will provide you with the apparatus that you may need for your investigation.

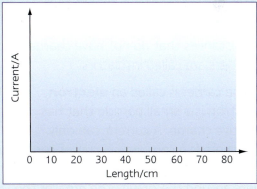

power pack / ammeter / bare wire / heatproof mat / crocodile clip

Method:

1 Set up the circuit as shown in the diagram. The wire can get hot. Take care not to burn your fingers.

2 Clip the crocodile clips on to the wire at a distance of 10 cm apart.

3 Copy the table below into your notebook to record your results.

4 Switch on the power pack and read the ammeter. Record the current in your table. Switch off again.

5 Move the crocodile clips until they are at a distance of 20 cm apart, and then repeat step **4**.

6 Keep moving the crocodile clips 10 cm further apart each time and measuring the current until you have completed your table.

7 Copy the graph axes below into your notebook and use your results table to draw a line graph.

Length of wire (cm)	Current (A)
10	
20	
30	
40	
50	
60	
70	
80	

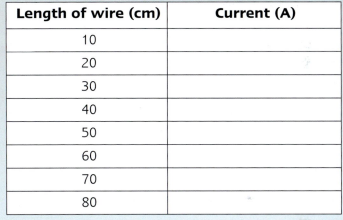

Assess Yourself

1 Explain in your own words what you think your graph shows.

2 What length of wire has the highest resistance?

3 Why do you think this wire has the highest resistance?

4 Are there any results that do not seem to fit on the line? If so, which ones are they? Can you think why this might have happened?

Modelling circuits

BIG IDEAS

You are learning to:
- Describe how electrical circuits work
- Recognise the relationship between current and voltage
- Describe resistance

Why do we use models?

Understanding electrical circuits can be difficult. This is because there are three **variables** to consider. These are:

- the current
- the voltage
- the resistance.

We cannot see any of them! Sometimes it is helpful to use a **model** to help us understand something. It is important to remember that no model is perfect.

Current

Current is carried by **charge carriers**.

- Substances that have charge carriers are called conductors.
- Substances that do not have charge carriers are called insulators.

A charge carrier is called an **electron**. An electron is a small particle that has a negative charge. A current can only flow if there is a complete circuit of conducting material and something to push the charge carriers around the circuit. The greater the number of charge carriers that move around the circuit the higher the current.

a
conducting wire   charge carriers in the wire (electrons)

b
energy source (eg battery or power supply)

complete circuit of conducting material

c
low current

high current

FIGURE 1: Models to show the relationship between current and voltage in an electrical circuit. In the diagrams what is used to model the electrons? And what is used to model the 'push' from a battery?

Current (I) is the movement of electric charge around a complete circuit of conducting material. It is measured in amps, A.

1 What are the **three** variables in an electrical circuit?

2 Why is it useful to use a model to explain what is happening in an electrical circuit?

3 What carries current around a circuit?

4 What is the difference between a high current and a low current?

Watch Out!

Remember that current doesn't get used up around a circuit – it keeps going around the whole circuit.

... charge carrier ... current (I) ... electron ... force ... heating effect

Voltage

The **force** that is pushing the current around a circuit is called the **voltage**.

> Voltage (V) is a measure of the energy supplied by a battery or power supply to move current around a circuit. It is measured in volts, V.

The higher the voltage the more energy is provided to the circuit.

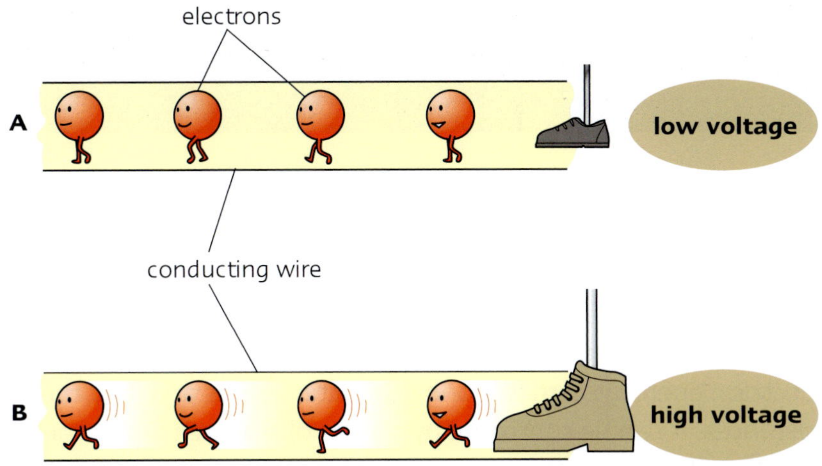

The energy that is provided to a circuit is transferred by the components in the circuit into different forms of energy. This transfer of energy allows the components in the circuit to work.

electrons

A   low voltage

conducting wire

B   high voltage

FIGURE 2: Using the same model to explain how changing the voltage makes electrons behave in different ways in an electrical circuit.

5 What is voltage a measure of?

6 What is the difference between a high voltage and a low voltage?

7 What happens to the energy in a circuit when the current passes through a component?

Resistance

The higher the resistance in a circuit the more energy is needed to move the current through it. So, to move a high current through a high resistance requires a high voltage.

> **Resistance (R)** is the difficulty that charge has in moving around a circuit. It is measured in ohms, Ω.

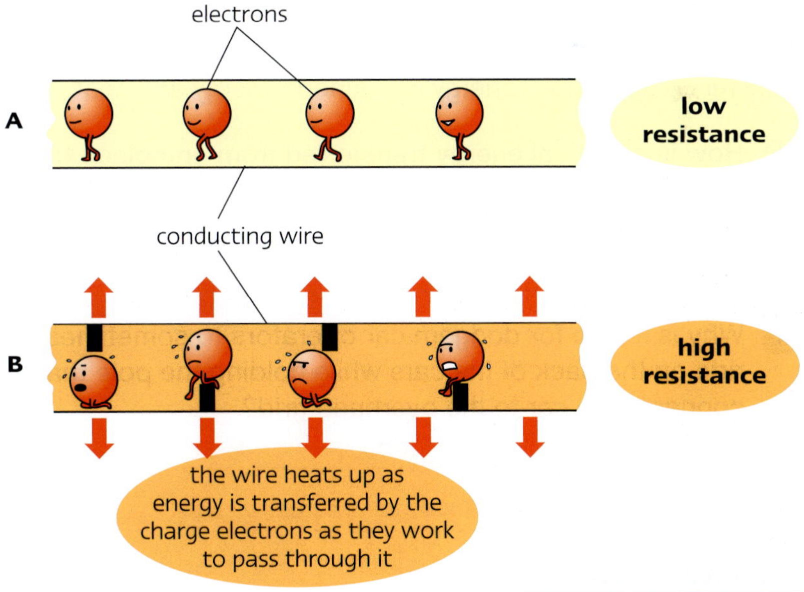

When the resistance is very high a lot of energy is transferred as heat into the wire or the component. This is called the **heating effect** of a current.

Sometimes the heating effect is useful and sometimes it is a problem.

electrons

A   low resistance

conducting wire

B   high resistance

the wire heats up as energy is transferred by the charge electrons as they work to pass through it

Using the same model to explain the heating effect in an electrical circuit. Why is a lot of energy transferred as heat when the resistance of a wire is high?

8 Why is a high voltage needed to move a high current through a high resistance?

Using electricity

BIG IDEAS

You are learning to:
- Describe some of the advantages of using electrical energy
- Explain some of the uses of electrical energy
- Evaluate uses and benefits of scientific developments

Transferring electrical energy

Electrical energy can be easily **transferred** from one place to another by means of an electric circuit. The electrical energy can come from the **mains** supply or from the chemical reactions in **batteries**. Electrical energy is carried along conducting wires and it is, therefore, possible to move it over long distances. It can be controlled by means of a switch which means it can be rapidly switched on or off and therefore there is a minimum amount of waste.

By altering the current, voltage and resistances within any circuit, the amount of energy that can be transferred from one place to another can be easily controlled.

Flood lights at sports stadiums transfer electrical energy from the national grid into heat and light to enable sports to be played in poor light conditions.

FIGURE 1: Electrical energy in use at a sports stadium.

Dodgem cars

Dodgem cars use an electric current flowing in an overhead grid to power an electric motor. The current flowing in the overhead grid is controlled by the operator and the motor transfers the electrical energy into kinetic energy and heat.

1 How is electrical energy transferred from one place to another?

2 Explain why electricity is such a useful and convenient form of energy.

3 Why is it safe for dodgem car operators to sometimes ride on the back of the cars while holding the pole that connects the car to the overhead grid?

FIGURE 2: How do dodgem cars use electrical energy?

Medical applications of electricity

Mobility scooters

Electric scooters, like all electric vehicles, rely on a rechargeable battery to provide the energy source. An electric current is then used to transfer the energy from the battery to an electric motor which in turn transfers it into kinetic energy which can be used to drive the vehicle.

Infrared heat lamps

Infrared heat lamps transfer low amounts of electrical energy into heat. Their filaments work at a lower temperature than normal light bulbs, so they emit much less light and more infrared radiation.

... batteries ... capillaries ... defibrillator ... infrared

Instead of heating the air around the body, the infrared (IR) energy is transferred to the cells underneath the skin, heating them up. This allows infrared heat lamps to push toxins and sweat from the body and cause the dilation of **capillaries** which improves circulation. With an infrared heat lamp, problem areas can be targeted. For example, those with poor circulation to their ankles, feet and toes may find the use of a targeted infrared heat lamp on their feet will increase the circulation and eliminate swelling and numbness.

FIGURE 3: What are some uses of heat lamps?

4 Explain why an electric car needs to be recharged regularly.

5 What would happen to the filament in an IR lamp if too much current flowed into it?

6 Why would heating the skin cause capillaries to dilate?

Defibrillators and pacemakers

Defibrillators

A **defibrillator** is an electrical device used to treat people who have suffered a heart attack. A defibrillator works by placing two paddles on the patient's chest and then transferring an electric current through the chest cavity to the heart where it discharges and hopefully causes the heart to start contracting normally.

Pacemakers

When your heart beats, an electrical signal spreads from the top of the heart to the bottom. As it travels, this electrical signal causes the heart to contract. Sometimes this does not work correctly and an artificial **pacemaker** is fitted. A pacemaker is a small device consisting of a battery, a computerised generator, and wires with electrodes on one end. The battery powers the generator and the wires connect the generator to the heart. It is placed under the skin of your chest or abdomen to help control abnormal heart beat.

The pacemaker's generator produces electrical pulses which are sent to and from your heart. Pacemakers have one to three wires that are each placed in different chambers of the heart. The electrical energy is transferred to the heart tissue causing it to contract.

A computer chip uses the information it receives from the wires connected to the heart to ensure that the right current is sent at the right time to the right part of the heart.

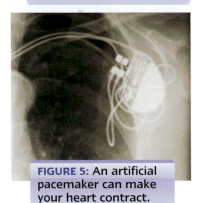

FIGURE 4: An electrical device that saves lives – the defibrillator.

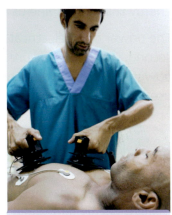

FIGURE 5: An artificial pacemaker can make your heart contract.

7 Many emergency services carry defibrillators and some are found in public buildings. Explain why.

8 Why are some pacemakers fitted internally while some are carried externally by the patient?

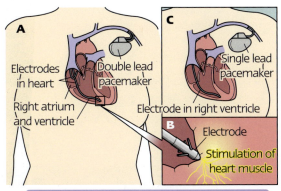

A Electrodes in heart Double lead pacemaker Right atrium and ventricle

C Single lead pacemaker Electrode in right ventricle

B Electrode Stimulation of heart muscle

FIGURE 6: How a pacemaker works.

Electricity in the home

BIG IDEAS

You are learning to:
- Describe how electricity is applied in the home
- Describe how improving the efficiency of appliances reduces energy consumptions and increases the sustainability of energy resources
- Explain the idea of a carbon footprint

Domestic electrical appliances

Many of the common appliances that are used in the home work by transferring electrical energy into different forms of energy. The amount of energy that a machine uses each second is called its **power**.

The power of an electrical device tells us how much energy it transfers in one second. Power is measured in **watts** (W) or kilowatts (kW). The higher the power of the appliance, the more energy it uses. The amount of energy that is used by the appliance depends on its power and the time it is working for.

How Science Works

Energy efficiency
More efficient machines are better at converting input energy into useful output energy. The total amount of energy before and after any energy transfer is always the same.

How good a machine or device is at usefully transferring energy is called the **efficiency** of the machine. This figure is usually given as a percentage. If a cooker is 75% efficient, it means that for every 100 J of electrical energy it is provided with, it transfers 75 J into useful heat output. The remaining 25 J are transferred as non-useful light energy.

The efficiency can be calculated in the following way:

efficiency = useful energy output ÷ total energy input x 100%

So for the cooker:
efficiency = (75 J ÷ 100 J) x 100% = 75% efficient
To be able to work out the efficiency of any energy transfer device we must be able to recognise what the useful energy output is in each case. This will depend on what it is that we want the machine to do.

Calculate the efficiency of a traditional 60 W light bulb which transfers 55 J of every 100 J of energy supplied into heat and an energy-efficient 11 W light bulb which transfers 90 J of every 100 J into light. What happens to the energy that is not usefully transferred in each case?

... carbon footprint ... efficient ... efficiency ... energy rating

Sustainable energy use

As energy supplies have become more scarce, the price of energy has increased. This has led manufacturers to develop more **efficient** systems that use less energy and that increase the **sustainability** of dwindling energy resources. Most modern electrical appliances used at home now have an **energy rating**.

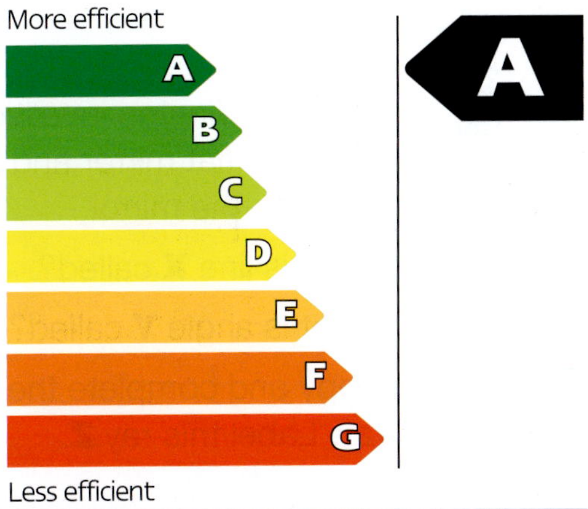

The EU energy label rates products from A (the most efficient) to G (the least efficient). For refrigeration the EU energy label goes up to A++. By law, the label must be shown on all refrigeration appliances, electric tumble dryers, washing machines, washer dryers, dishwashers, electric ovens, air conditioners, lamp and light bulb packaging. By choosing more efficient appliances, people can help make energy resources more sustainable.

More efficient

A
B
C
D
E
F
G

A

Less efficient

FIGURE 1: The EU energy efficiency rating.

1 Why have manufacturers of electrical appliances made them more energy efficient?

2 What advantages are there in providing energy ratings for electrical appliances?

Carbon Footprint

The term '**carbon footprint**' is becoming increasingly common. The carbon footprint is the total amount of greenhouse gas emissions caused directly and indirectly by an individual, event, organisation or product (in its manufacture and usage). Electricity is commonly generated by burning fossil fuels and this releases carbon dioxide gas into the atmosphere which in turn contributes to global warming.

The UK's carbon footprint is over 500 million tonnes of CO_2 per year. Individuals account for 45% of this from their use of lighting, heating, for example, and for the powering of equipment. By reducing their carbon footprint people can lessen their contribution to global warming. One way to do this is to only use electrical appliances when it is necessary. This has the added benefit of making energy resources more sustainable.

3 Why is the carbon footprint of a country becoming more important?

4 Explain the link between the carbon footprint of a product and its energy efficiency.

5 What issues will need to be addressed in order to reduce the carbon footprint of a country? Explain your answer.

1 Tracey and Winston investigate reflection. They shine a beam of
light on to a flat mirror and notice what happens to the beam
when it hits the mirror.

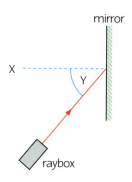

a What is line **X** called?

b What is angle **Y** called?

c Copy and complete the diagram above to show the reflected
ray. Label this ray **Z**.

2 Choose the correct statement.

a A red book appears red because it absorbs red light.

b A red book appears red because it reflects red light.

c A red book appears red because it transmits red light.

d A red book appears red because it emits red light.

3 When white light is shone at a glass prism the light splits
up into the colours of the spectrum.

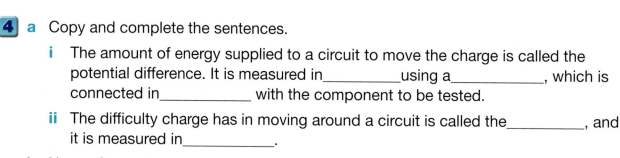

a Copy and complete the diagram to show what
happens when the light passes through the prism.

b What are the colours of the spectrum?

c On your diagram label the colours as they appear when they emerge from the prism.

4 a Copy and complete the sentences.

  i   The amount of energy supplied to a circuit to move the charge is called the
      potential difference. It is measured in_____using a_____, which is
      connected in_____ with the component to be tested.

  ii The difficulty charge has in moving around a circuit is called the_____, and
      it is measured in_____.

b Name **three** things that increase the resistance of a wire.

5 Copy and complete each of the following sentences using only the words 'increases'
and 'decreases'.

a For a constant voltage, if the resistance is increased the current_____. If the
current is increased the resistance_____ .

b For a constant current, if the voltage is increased the resistance_____. If the
voltage is decreased the resistance_____ .

c For a constant resistance, if the current is increased the voltage_____.
If the current is decreased the voltage_____ .

6 Coloured filters are useful because they only let certain colours of light pass through. Copy and complete the diagrams below to show what colours of light pass out of each filter.

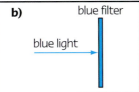

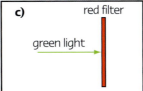

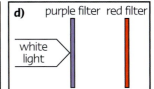

7 Write down each part of the eye and its correct function.

Part:	Function:
lens	light is focused on to this which acts as a screen
cornea	hole in the eye that lets light in
optic nerve	clear transparent window at the front of the eye
pupil	coloured part of the eye
retina	focuses the light
iris	carries a signal to the brain

8 Choose the correct statement.

   a Red light refracts most in a prism because it has the shortest wavelength.

   b Red light refracts most in a prism because it has the longest wavelength.

   c Violet light refracts most in a prism because it has the longest wavelength.

   d Violet light refracts most in a prism because it has the shortest wavelength.

9 Copy and complete the table below to show the colour an object appears when it is put in differently coloured light.

Colour of object	In white light	In green light	In red light	In blue light
red				
blue				
green				
yellow				
purple				
black				

10 Here are some angles of incidence (i) for a light wave hitting a glass block. For each one the angle of refraction is given.

Angle of incidence (i)	0	10	20	30	40	50	60	70	80	90
Angle of refraction (r)	0	7	13	19	25	30	35	39	41	42

   a Explain why the light ray refracts when it enters the block at an angle.

   b Plot a graph of angle i against angle r.

   c What angle of incidence gives an angle of refraction of 10°?

   d What is the angle of refraction if the angle of incidence is 36°?

Topic Summary

Learning Checklist

☆ I can explain how shadows form.   page 132

☆ I know that light travels much faster than sound.   page 132

☆ I can describe how light is reflected at plane (flat) surfaces.   page 134

☆ I know that filters and coloured objects absorb some colours (and transmit or reflect others).   page 144

☆ I know that light travels from light sources to our eyes so we can see them.   page 148

☆ I know that most insects have compound eyes.   page 148

☆ I know how to measure the current in a circuit.   page 152

☆ I can use the terms current, voltage and resistance in my explanations.   pages 152, 154

☆ I can explain **two** ways of changing the size of a shadow.   page 132

☆ I can describe how light is refracted at plane (flat) surfaces.   page 134

☆ I can describe how curved mirrors work.   page 136

☆ I can describe how a prism affects white light.   page 142

☆ I can describe what effect coloured filters and coloured light have on coloured objects.   page 144

☆ I know that we see objects' reflected light.   page 144

☆ I can give **two** examples of how colour is useful in everyday life.   page 146

☆ I know the main parts of the eye and can label them on a diagram.   page 148

☆ I can describe what electrical resistance is.   page 154

☆ I know what affects the resistance in a circuit.   page 154

☆ I can demonstrate the Law of reflection at plane (flat) surfaces.   page 134

☆ I can describe total internal reflection.   page 138

☆ I can recognise refraction and dispersion and I can give examples of each.   pages 140, 142

☆ I can explain how the eye works.   page 148

☆ I know the relationship between the resistance of a component and the current it allows to flow.   page 154

☆ I know how convex lenses and concave lenses work.   page 136

☆ I can explain why coloured objects look different in coloured lights.   page 144

☆ I know that how much light refracts depends on its colour.   page 144

☆ I can explain the behaviour of circuits using a charge flow model.   page 154

☆ I can describe the different parts of the Electromagnetic Spectrum.   page 143

Topic Quiz

1 Name **two** materials light can travel through.

2 What happens when light hits a flat surface?

3 What is an image?

4 What does 'plane' mean?

5 What is dispersion?

6 List the colours of the spectrum.

7 What does a coloured light filter do?

8 Name **two** things that emit light.

9 What does a prism do to white light?

10 What happens to a ray of light when it enters a denser material?

11 What colour is a red object in red light?

12 **a** What colour is a red object in green light? **b** Explain your answer.

13 How can you make your shadow smaller?

14 State the Law of reflection.

15 What does the retina do?

16 **a** Would blue light pass through a blue filter? **b** Would blue light pass through a red filter? **c** Explain your answers.

17 What is electrical resistance?

18 What **three** things affect the resistance of a wire?

19 What does voltage measure?

20 What does current measure?

True or False?

If a statement is incorrect then rewrite it so it is correct.

1 Light travels faster than sound.

2 Bright colours are used for camouflage.

3 Black objects reflect all the light that hits them.

4 Shadows form because light travels through an object.

5 Reflection happens when light bounces off a flat surface.

6 Red light and green light combine to make yellow light.

7 Green light and blue light combine to make magenta light.

8 Red light has a longer wavelength than violet light.

9 The resistance decreases when extra bulbs are put in the circuit.

10 A small voltage is needed to make a large current flow through a high resistance.

11 Most components have resistance.

12 Another term for voltage is potential difference.

13 The power of an appliance measures how quickly it uses energy.

ICT Activity

Choose **one** of the following three questions. When you have made your choice carry out some research to find out information so that you can answer the question. Present your answer as a PowerPoint slideshow.

1 Find out how the design of sight glasses has changed in the last 500 years. What developments have helped this to happen?

2 What material are contact lenses made from? How many different types of contact lens are there? What are these different types used for?

3 How are lasers used in the treatment of detached retinas?

How are living things coping with global warming?

The polar bear relies on ice for rearing its young and also for hunting. Its main source of food is the seal. The seal also relies on ice to rear its young. It is reported that some of the ice cover on the Earth is beginning to melt because of global warming and this will affect polar bears and seals. There is evidence that polar bears weigh less than they did 20 years ago.

Why is global warming happening?

The world is the warmest it has been for over 12 000 years. Some scientists say the Earth has warmed by about 0.2 °C in the last 30 years.

One of the effects of global warming is the change it causes to animals' and plants' habitats. If there is less snowfall in winter, warmer temperatures in the summer and more rainfall the behaviour of animals is affected. For example they can migrate to new areas to find new food sources – this is one way of coping with climate change. Plants cannot migrate even if their seeds and fruits can relocate.

However for some animal species migrating to cooler regions is a struggle. The Worldwide Fund for Nature states that by 2010 global warming may have driven some species to extinction.

A polar bear and its seal kill.

HUMANS AND THE ENVIRONMENT

- The worst affected areas are the higher northern latitudes that have lost up to 70% of their habitat.
- Coastal areas are also suffering from a combination of rising sea levels and warming of the oceans.

A recent nature study reported that, 'birds are laying eggs earlier than usual, plants are flowering earlier and mammals are breaking hibernation sooner'.

Among the changes recorded in the study were:

- birds such as the ptarmigan that are found in Britain cannot survive in warmer conditions, so they are moving further north, away from Britain
- many fish species are moving northwards in search of cooler waters
- marmots are ending their hibernation 3 weeks early
- the golden toad of Costa Rica is now extinct and is thought to be one of the first victims of global warming.

BIG IDEAS

By the end of this unit you will be able to explain the impact that humans have had on the environment and evaluate some of the strategies needed to conserve the environment. You will have used various sources of information and explored a variety of responses.

The Asian tiger mosquito carries diseases not normally found in Britain. In the summer of 2007 it was spotted in Britain.

Insect invasion!

Mosquitoes that carry diseases not normally found in Britain, for example malaria and dengue fever, are likely to spread dramatically because of global warming. Dengue fever is a severe flu-like illness that often causes fatal internal bleeding and affects over 50 million people in the tropics.

Global warming has resulted in increased variability of the climate – warmer winters and hot, dry summers ideal for the mosquito – so they have thrived in countries where before the climate has been unsuitable. This caused the surprise occurrence of West Nile disease in New York City in 1999. The mild winter enabled many mosquitoes to survive through to spring. The drought in the spring killed many of the mosquito's predators and the mosquitoes in the area acquired the disease. The torrential rains in August then provided ideal conditions for the mosquito to breed. This resulted in the rapid spread of the disease.

What do you know?

1 What is the polar bear's main source of food?

2 Why are there fewer seals for the polar bear to eat?

3 Suggest why might polar bears weigh less than they did.

4 What other changes in animal behaviour are happening because of global warming?

5 Why does spring seem to be occurring earlier?

6 Why are animals migrating to the cooler areas?

7 Why are plants not migrating?

8 What problems do animals face when they leave their normal habitats?

9 Why might malaria return to Britain?

10 Why was there an outbreak of West Nile disease in New York?

What resources do we need?

BIG IDEAS

You are learning to:
- Describe the effects of land and water pollution on the environment
- Discuss the problems caused by an increased population size

Pollution of land

Land is needed for farming to make food, for housing, industry, mining and transport. Much of the land on Earth is unusable, for example deserts and mountains. Therefore, as **populations** grow, natural grassland and forests are removed to provide more usable land. The removal of forests (**deforestation**) can lead to soil erosion and the loss of habitats for wildlife.

Humans' development of the land produces **pollution**. The more land that is used by humans the more of it becomes polluted.

FIGURE 1: 'Slashing and burning'. Part of this rainforest is being cleared to make way for a new road. Do you think the benefits of clearing forest outweigh the environmental damage caused?

1 Why is more land needed by humans?

2 What problems are caused by deforestation?

Pollution of water

Clean water is needed for drinking, preparing food and washing in. Most of the water on Earth is in the sea and is not fit for drinking. Rivers can become polluted by raw sewage, fertilisers and pesticides washed off from farmland and waste materials from mining and industry. The seas can also be polluted by oil spillages from tankers.

Did You Know...?

Sewage consists of human waste, household waste and run-off from streets. Snakes and alligators that have grown too big for their owners have been washed down the sewers in the USA.
They survive by feeding on rats that live in the sewers.

How Science Works

A sewage works needs bacteria. The sewage is first left to settle out and then any small particles of sewage left in the water are broken down by bacteria in controlled conditions so that the water can be returned to the river.
The sludge left after settling is broken down without oxygen being available to make methane gas. The gas is burnt to make electricity.

3 How do fertilisers end up in the sea?

4 What problems could be caused by sewage being directly released into the sea?

... deforestation ... developed ... environmental damage

Environmental damage

The World has a growing population of over 6 billion people. As the number of people on Earth increases so do pollution and **environmental damage**. About 25% of the world's population are causing 75% of the damage to the environment. The **developed** countries use many of the World's resources. As these resources are processed pollution is produced.

- Oil is used for making fuel for transport, plastics and generating electricity.
- Wood is used for making paper and furniture.
- Mining provides ores from which metals are extracted, precious gems and coal for making electricity.

Mining also leaves scars on the landscape. This is an example of environmental damage. Other examples are over-fishing the seas so that there are no longer enough fish to catch and farming **intensively** (producing more food from the same area of land). Diseases and pests become more of a problem in intensive farming because large numbers of the same animals or plants are raised/grown. Therefore farmers have to use more pesticides, fertilisers and medicines. To get the most out of their land farmers who use intensive farming sometimes cut down the hedgerows.

5 Explain why pesticides and fertilisers are used more in intensive farming.

6 Suggest how the removal of hedgerows can lead to an increase in pollution.

FIGURE 2: Intensive farming in America. What features in the photograph tell you the farming method is intensive? Can you list the advantages and disadvantages of this type of farming?

A new by-pass

Case study

A new by-pass is to be created round Steeltown.

The following views have been expressed.

Local resident
The heavy traffic is damaging the road, causing high noise levels and air pollution. It causes long delays.

Council advisor
The cost is very high and will cause local tax increases.

Golfer
It will ruin the best golf club in the area.

Local industry
Congestion stops lorries getting quickly to our factories.

Environmentalist
The woodland area has many very rare plants. They cannot be moved anywhere.

Farmer
Farmland will be used up.

You are a member of the council. Suggest what evidence you will need to collect to make your decisions.

7 Decide whether you feel the by-pass is justified. Give a reasoned explanation of your decision.

The effect of acid on plants

BIG IDEAS
You are learning to:
- Explain the causes of acid rain
- Describe the effects of acid rain on plants
- Evaluate the evidence for the effect of acidity on plants

Where does acidity come from?

Normal rain is slightly **acidic** (pH 5.6) because it has some carbon dioxide dissolved in it. The main gases produced by vehicles and by industry are **sulphur dioxide** and **oxides of nitrogen**. When these **dissolve** in rain they make it more acidic (about pH 4.0).

1 Explain why the rain in the following areas has the pH measured.

Area	pH
Industrial area	4.1
National park	5.6
Congested city	4.5

Effects of acid rain

Acid rain affects plants by:

- washing important minerals out of the soil so plants become short of minerals
- removing the waxy outer layer of plant leaves. This damages the leaves so photosynthesis cannot take place – the leaves turn yellow and eventually drop off
- not allowing plant cells to develop properly.

The leaf on the top has been affected by acid rain and its cells have not developed properly; the leaf on the bottom is healthy. Can you see the differences in the shapes and number of the cells in the two pictures?

2 **a** Describe the differences in the shapes and number of cells.
   b Explain how this affects the growth of the plant.

3 How does acid rain cause plants to lose their leaves?

4 Why do plants need a waxy layer on their leaves?

Did You Know...?

Acid rain draining into freshwater lakes causes the release of heavy metals, for example mercury, aluminium and cadmium and these enter the food chain. Aluminium is deposited in the gills of fish and builds up in their tissues.

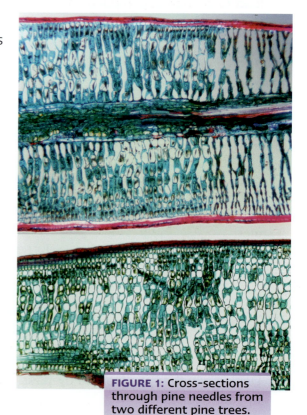

FIGURE 1: Cross-sections through pine needles from two different pine trees.

... acid rain ... acidic ... dissolve ... oxides of nitrogen

Can seeds grow in acidic conditions?

Cress seeds are easy to germinate and look after so they make good seedlings to use in this investigation.

You are going to investigate how different concentrations of acid effect the germination of seeds.

Your teacher will provide you with the apparatus that you may need for your investigation.

Method:

1   Place the filter paper in the dish and add 5 cm³ of water to the dish.

2   Sprinkle 10 cress seeds evenly over the filter paper. Put the lid on and write 'water' on it using the marker pen.

3   Take another dish and repeat steps **1** and **2** and add 1.0cm³ acid. Write on the lid '1.0cm³ acid'.

4   Repeat steps **1** and **2** adding 2.0 cm³, 3.0 cm³ and 5.0 cm³ acid. Remember to label each of your dishes correctly.

5   Leave the seeds to germinate for 1 week and record your observations in your notebook.

Cress seedlings.

Questions

1   How was the test made fair?

2   What pattern do your results show?

3   Why did you cover the dishes?

4   Why did one of the dishes have only water in?

5   Design an experiment to find the effects of different levels of acidity on the growth of seedlings, not seeds.

6   How do your results compare to those of other groups? Evaluate your method and the validity of your data based on this comparison.

How clean is our air?

BIG IDEAS

You are learning to:
- Describe the effect of soot on an organism
- Explain how soot is caused and how soot emissions are controlled
- Examine some evidence for the levels of soot pollution in the area

HSW

What is soot?

Soot is **carbon** that has not burnt properly. Soot is made of fine **particles** and forms a layer that covers the leaves of plants and trees and their trunks. Soot particles are produced when **fuels** are burnt. Coal fires produce high levels of soot.

1 Why were some of the buildings in the 19th Century turned black?

2 Why is there less soot in our cities in this century?

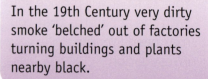

FIGURE 1: Soot particles are made when fuels are burnt. They pollute the surrounding area.

Did You Know...?

In the 19th Century very dirty smoke 'belched' out of factories turning buildings and plants nearby black.

FIGURE 2: A church in London partly covered by soot.

What effects does soot have?

- Soot coats the surface of a leaf causing less light to reach the leaf. This means that less photosynthesis occurs. Soot also contains chemicals that destroy the waxy layer on leaves.

- Humans and other animals are affected when they breathe in soot. Their lungs are damaged leading to an increased risk of bronchitis and lung cancer.

3 Why does a plant grow less well if it is covered in soot?

4 How do soot particles cause bronchitis in animals and people?

How Science Works

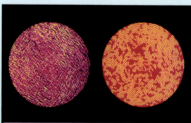

On the left is a soot particle seen under a very powerful microscope called an electron microscope and on the right is its 'fingerprint' image created on a computer.

Scientists have worked out a way of 'fingerprinting' soot particles. For example they can tell if a soot particle has come from burnt wood or coal or from a diesel engine. Scientists hope to use these soot 'fingerprints' to identify the sources of soot pollution.

HSW

Black soot causes Arctic melting

5 Explain why the soot will cause the ice and snow to melt.

6 Suggest what effect this will have on the environment.

Case study

Researchers at University College Irvine estimate that 33% of Arctic warming is caused by soot mixed in the snow and ice. The soot is blown in the wind from China's coal plants and Asian fires and causes the melting of the snow and ice due to solar radiation falling on it.

... carbon ... fuel

Investigating soot levels in the environment

The leaves of plants that grow in different areas can be picked and examined to see the amount of soot covering them. In this way pollution levels in different areas can be compared.

You are going to investigate the amount of soot in two different areas. The pine needles are all from trees of the same age.

Your teacher will provide you with the apparatus that you may need for your investigation.

The pine tree has suitable leaves for this type of study.

Method:

1 Cut the paper into two 4 cm squares and then fold each square in half.

2 Take 10 pine needles of similar length from trees growing in the first area.

3 Place one needle in the folded paper and pull it through pressing gently on the needle.

4 Repeat step **3** using the *same* piece of paper pulling the remaining nine needles through.

5 Now use the second piece of paper and draw through the 10 needles from trees growing in the second area.

6 Stick the two squares in your notebook. Remember to label the squares!

Questions

1 How was your test made fair?

2 What caused the black marks on the paper squares?

3 What do your results show?

4 Suggest why needles of the same age were used in the investigation.

5 Evaluate the results collected and the methods used to collect the results.

Councils introduce rubbish recycling and fortnightly collections

To make better use of our resources and reduce the amount of waste taken to landfill sites councils have introduced recycling. Differently coloured plastic boxes are provided for certain groups of waste and this is collected separately and then recycled. Materials put in these recycling boxes are glass, paper, cardboard, metal and, in some cases, plastic and garden wastes. By recycling material less of our natural resources will be used up.

The rubbish in wheelie bins is emptied by some councils once a week but more and more councils are changing to a fortnightly rubbish collection. The waste is taken to a landfill site to be compressed and then buried. This does not get rid of the material as very little of it decays. Landfill sites only hide waste and also cause problems for the future such as the pollution of nearby soil. Suitable areas for landfill sites are running out and it costs councils money to dump rubbish at existing sites.

Rubbish from wheelie bins is dumped at landfill sites. We are running out of suitable places to have these sites.

So, what are the advantages of recycling?

- Paper is pulped and used to make recycled paper – therefore fewer trees have to be cut down.
- Glass is ground up and used as a starting material to make new glass.
- Using magnets metal cans can be separated into aluminium and steel. The metal is then melted down and reused.

Is fortnightly often enough?

- People are worried that by collecting rubbish only every two weeks councils will encourage illegal dumping of excess waste material that damages the environment.
- There is also concern that during the summer months the smell of rotting waste will attract rats, flies and cockroaches and make it very unpleasant for people in their houses. Rats are known to carry diseases.

Illegal dumping. It is unsightly and damages the environment.

Rats carry diseases.

It is unlikely that conditions will get as bad as the 16th Century when the Black Death transmitted by rat fleas killed thousands of people. Conditions for the spread of disease will be present, however, and therefore good hygiene will be essential.

Councils have indicated that by emptying bins every two weeks they can meet their recycling targets laid down by the Council of Europe. They also find it is more economical to collect the rubbish in this way as less goes to landfill sites and more is recycled.

Assess Yourself

1. Give **three** problems caused by leaving rubbish in the streets for long periods of time.

2. a Name **three** materials that are recycled.

   b Explain what happens to each material.

   c What are the benefits of recycling?

3. a What are landfill sites?

   b What are the problems of using landfill sites i) short-term ii) long-term?

4. How does a council obtain its money?

5. a Explain the problems caused by the council spending more money on waste disposal.

   b What could be the short and long-term effects of raising taxes on local industry to make them reduce pollution?

6. Produce a paper for a council meeting which shows the case **for** and **against** the emptying of bins every two weeks (approximately 150 words max).

7. Plastic and polystyrene are causing many problems. Present an argument **for** and **against** the use of plastic packaging and containers. Include:
   - the problems they cause
   - possible solutions to the problems including alternative forms of packaging
   - why plastic has been used.

Mathematics

Count the number of tin cans, bottles and cardboard packages disposed of in one waste collection cycle in your household. Draw a bar chart or a pie diagram to display your results. Which waste type do you recycle most?

Design and technology

Design a simple method to help pensioners with walking problems carry their plastic recycling boxes out of their houses ready for collection.

Level Booster

EP Your answers are able to describe and explain the application and implications of using packaging material and its effect on the environment.

8 You are able to produce a reasoned argument relating to the case for and against the collecting of rubbish on a two weekly cycle.

7 Your answers explain the implications on waste disposal of making economic decisions.

6 Your answers show you can apply your knowledge and understanding of using landfill sites to consider the short and long term effects.

5 Your answers describe the benefits of recycling.

4 Your answers describe the implications of allowing rubbish to be left.

What happened to the atmosphere?

BIG IDEAS

You are learning to:
- Describe the causes of different types of air pollution
- Explain the effect of the types of air pollution on the environment
- Discuss the effect of technology on the environment

Pollution of the air

When fossil fuels (oil, coal and gas) are burned they **pollute** the air with smoke particles and chemicals called **carbon dioxide**, **carbon monoxide** and **sulphur dioxide**.

Cars, lorries, buses and industry, such as power stations, all use fossil fuels. They are the main causes of pollution.

FIGURE 1: Pollution from vehicles is a big problem in the World.

1 Suggest why policemen controlling traffic in busy cities suffer from respiratory problems.

2 Why are large cars taxed more than small cars?

Smog

Smog is formed when **smoke** particles mix with **fog**. When smoke particles are breathed in they can cause damage to the lungs. Smog is still a major problem in large cities in developing countries. In these countries cities and industries have grown at such a speed that the number of coal-fired power stations has also risen dramatically.

3 How did the introduction of smokeless zones (areas not using coal for fires) in London in the 1950s help to reduce smog?

4 Describe **three** ways Shanghai could reduce its smog.

Did You Know...?

Smog was a major problem in the 1950s in London and other industrial cities due to the use of coal as a fuel in peoples' homes. Smog was also called a 'pea-souper' as it was greeny-yellow in colour and people could not see through it.

'Smokeless' fuels and smoke-free zones were introduced to reduce the smog. In these zones coal could not be burnt on fires in the home.

FIGURE 2: Smog lying over Shanghai in China. Why are developing countries experiencing smog now, while Britain does not in general have a smog problem?

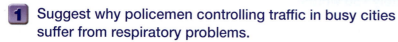

... acid rain ... carbon dioxide ... carbon monoxide ... CFCs ... global warming

Acid rain

Sulphur dioxide and **oxides of nitrogen** given off in smoke dissolve in water in clouds so that when rain falls it is actually a weak acid – called **acid rain**.

Acid rain can cause the following problems:

- breathing difficulties in people leading to shortness of breath and sometimes asthma
- damage to tree leaves, sometimes leading to death of the tree
- erosion damage of buildings made from limestone or marble.

5 Suggest **two** ways technology has caused air pollution and **two** ways it has reduced pollution.

6 Many countries produced very large chimneys for the removal of industrial gases. This removed the problem from their country but caused problems in other countries where the pollution 'fell'. Explain whether you think this is acceptable practice.

The greenhouse effect

Carbon dioxide is one of several gases in the atmosphere that are together called **greenhouse gases**. Carbon dioxide is produced when fossil fuels are burnt. Greenhouse gases act in a similar way to a greenhouse – they trap heat from the Sun. This warms the atmosphere which then causes the Earth to heat up.

As more carbon dioxide is produced, more heat is trapped and the Earth warms up even more. This effect is called **global warming**. It results in the melting of snow and ice on mountains and at the polar ice-caps. Many scientists agree that this causes flooding and changes in weather patterns.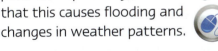

7 How can the government cut down carbon dioxide production?

8 Explain how our understanding of global warming has changed over the past 50 years. Indicate the investigations that have been carried out and the key evidence that has been gathered.

This ozone hole is above Antarctica (outlined in black) and is shown here by the pink, yellow and light-blue areas. The hole lets more ultra-violet radiation than normal into the Earth's atmosphere.

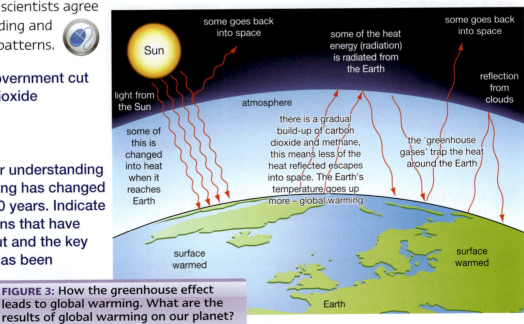

some goes back into space

some of the heat energy (radiation) is radiated from the Earth

some goes back into space

Sun

reflection from clouds

light from the Sun

atmosphere

some of this is changed into heat when it reaches Earth

there is a gradual build-up of carbon dioxide and methane, this means less of the heat reflected escapes into space. The Earth's temperature goes up more – global warming

the 'greenhouse gases' trap the heat around the Earth

surface warmed

surface warmed

Earth

FIGURE 3: How the greenhouse effect leads to global warming. What are the results of global warming on our planet?

... greenhouse gases ... oxides of nitrogen ... ozone ... pollute ... sulphur dioxide

Looking at your surroundings

BIG IDEAS

You are learning to:
- Explain how to measure the distribution of plants in an area.
- Explain how to measure the biodiversity of an area
- Analyse data on the distribution of plants

Studying the local environment

Case study – Studying the local environment

A field work study should start with a

- note of the date and time of day
- note of the location of the area and size of area
- study of the area to make sure you can identify most of the organisms found. This means the results can be checked by other people.

1 Explain why the results might be different at different times of year.

Variety and biodiversity

If you study a hedgerow there appears to be far more variety of plants than on a school field. This is because the plants are cut regularly on the field and many cannot grow. To compare the variety, a measure called biodiversity is used.

Biodiversity is the number of different **species** in an area and the **population** of each species.

- If an area has several plants but only one plant is common it has a low biodiversity.
- An area with a high biodiversity is rich in wildlife and is worth **conserving**.

FIGURE 1: Which of these environments has a high biodiversity and which has a low biodiversity?

2 A plantation of conifer trees has eight species of plants in it but only the conifer trees are common. Does it have a high or low biodiversity? Explain your answer.

3 Explain why a golf-green has a lower biodiversity than a football pitch.

4 Explain why the biodiversity of animals is higher in a hedgerow than on a football pitch.

How Science Works

Scientists can measure the amount of pollution in rivers by investigating the types and number of species living in the water.

This is because some organisms can tolerate higher levels of pollution than others. For example stonefly larvae are very sensitive to pollution, whereas sludgeworms can stand high levels of pollution. Scientists call these organisms **indicator species** because they can be used to indicate the levels of pollution in water courses.

Which of the species named above would you expect to find in high numbers in this highly polluted sludge lake?

... biodiversity ... conserve ... indicator species

Investigating the diversity of plants on the school playing field

Scientists can use different ways to measure the number of different species in an area. One of the pieces of apparatus they use is called a **quadrat**. It is a square frame that is placed on an area and the species inside it are counted and identified.

You are going to investigate how many different types of plants there are on your school playing field. You can use the plants shown below to help you identify the species growing in your playing field.

Your teacher will provide you with the apparatus that you may need for your investigation.

Method:

1   Select the area your group is going to study and mark out a 10-metre square.

2   Place your quadrat randomly in the square. (Draw two numbers between 1 and 10 from a bag and use them as grid references.)

3   For each quadrat area record in your notebook the different plant species found. (You may need to use reference books to identify a plant not shown above.) Use the following scale to record the abundance of each species.
   A abundant   **C** common   **R** rare

4   Repeat steps **2** and **3** for seven more different quadrat areas.

5   Find a very different area of land and repeat steps **1** to **4**.

6   Record your results in a table.

7   Work out the number of species found in each area from your results.

Questions

1   How did you place the quadrat at random?

2   Why is the quadrat placed at random?

3   Explain why the type of species and amount of each species varies between the area.

4   Which area is the most biodiverse?

5   Why is it difficult to count the number of plants in a quadrat?

6   Explain why some plants in the area were not recorded in the investigation.

7   Plan an investigation to find out on which side (N,S, E or W) of a tree the most green algae are found.

Conservation

Endangered species

Many animals are close to becoming **extinct** – they are called **endangered species**.

> A species becomes extinct when its numbers fall too low for it to replace animals that die.

Efforts are made to protect endangered species.

The causes of extinction of a species can be complicated but the main reasons are:

- hunting
- habitat destruction
- pollution.

FIGURE 1: In Britain the red squirrel and the osprey are protected.

How Science Works

Scientists are producing weights for fishermen that are not made from lead. Lead weights cause poisoning of birds such as swans that swallow them. The lead affects their nervous system.

HSW

Steps taken to protect species are:

- **habitats** are preserved
- trading bans are enforced
- the animals are captured and relocated
- **breeding programmes** are set up in zoos, wildlife parks and botanical gardens.

FIGURE 2: The giant panda, minke whale, gorilla and black rhino are endangered species that are protected. Do you know any other animals that are protected?

1 What methods can we use to preserve the gorilla?

2 Kew Gardens in London is a botanical gardens where a large number of plant species from around the world grow. Why is it important to grow different species of plants?

... breeding programme ... endangered species

Methods of conservation

More than 50 special wildlife areas are destroyed in Britain every year, as land is used for housing, roads and industry. Areas of land of special note are protected by the government. For example:

- setting up **green belts** (these are areas in cities where no building is allowed)
- national parks
- nature reserves
- sites of special scientific interest (SSSI).

Deciding which areas in a country are to be conserved is very difficult. Reasons to conserve an area may be:

- it is home to a very rare plant or animal
- it has a great diversity of wildlife
- it has outstanding natural beauty.

3 If the public wants to visit an area of outstanding natural beauty such as a waterfall how can the area be protected to prevent damage?

4 Do you think the presence of a rare plant should cause the re-direction of a planned motorway? Give reasons for your answer.

Why conserve?

Why should we worry about our surroundings? Would it make any difference if gorillas died out or green belt land was built on?

Some of the reasons that should be considered are:

- We have no right to destroy wildlife on the planet.
- The diversity of plants and animals may be needed in the future for medicines.
- Nature is wondrous and beautiful. Many people find they are able to find peace in the wild.

5 Present a reasoned argument for or against conservation of the great white shark.

6 There is a housing shortage. Give a reasoned case whether we should be allowed to build on green-belt land.

7 The rain forests of the Amazon in South America are being removed to develop farmland, housing, industry and for mining.
Present a case for or against this happening from the following peoples' viewpoint.
a A native who relies on the forest for housing or food.
b An industrialist.
c An environmentalist.
d A poor farmer.

Did You Know...?

The Millennium Seed Bank Project at Kew Gardens in London is the largest conservation project ever attempted.
The project is on course to bank 10% of the seeds of the world's wild plant species by the end of the decade. These will not be just any plants but will include the rarest, most threatened and most useful species known to man.

Storing the seeds in the bank. The seeds are kept dry and cool.

FIGURE 3: Part of the Snowdonia National Park. Do you think our country would be such a nice place to live if areas like this did not exist?

Can we save the planet?

BIG IDEAS

You are learning to:
- Locate information from different case studies
- Compare and contrast the effects of different causes of pollution
- Determine the key points for the introduction of fishing catches

The effect of the car

Road transport accounts for 22% of total UK emissions of carbon dioxide (CO_2) – the major contributor to climate change. Colour-coded labels, similar to those used on washing machines and fridges, are now displayed in car showrooms showing how much CO_2 new models emit per kilometre. However, as traffic levels are predicted to increase, road transport will continue to be a significant contributor to greenhouse gas emissions. Noise from road traffic affects 30% of people in the UK. Sources include engine noise, tyre noise, car horns, car stereos, door slamming, and squeaking brakes.

Did You Know...?

Health officers are paying increasing attention to sick building syndrome. The air pollution inside the office buildings can cause eye irritations, nausea, headaches, depression, fatigue and respiratory diseases.

Sometimes there is no choice and we just have to drive. But there are things you and your family can do to make sure your car pollutes less.

- First of all, buy the most fuel efficient car that meets your family's needs. This will lead to fewer emissions.
- Have a well-fitted fuel cap on your car to avoid the unnecessary release of fumes from your petrol tank into the air where they can be breathed in and contribute to photochemical fog.
- Think about converting to Liquid Petroleum Gas (LPG). This will not only reduce your contribution to air pollution, but is also more cost effective.
- Remove any extra wind resisting items from your car, such as the roof rack, you will save on fuel and at the same time improve air quality.
- Keep tyres inflated to the specified pressure to reduce resistance.

1 How does reducing wind resistance reduce pollution?

2 What benefit is there in using colour-coded labels for cars?

Flying to pollution

Aircraft account for about 5% of carbon dioxide emissions and air travel is forecast to double within 25 years. Besides carbon dioxide, jet engines emit many pollutants into the atmosphere, including oxides of nitrogen, sulphur dioxide, soot and even water vapor. Also, jet contrails – the vapor trails they leave in the sky – add to cloud cover and may contribute to the warming of the planet.

Noise from planes flying over residential areas impairs people's ability to work, learn in school, sleep, and consequently also results in lowered property values in affected areas. Passenger volume is increasing and new and larger planes are being produced.

Air transport is currently contributing around 3.5% to total human caused global warming but is forecast by the IPCC (Intergovernmental Panel on Climate Change) to rise to as much as 15% by 2050.

3 Compare and contrast the effect of pollution from cars and planes.

4 It has been suggested that airships could be used to reduce pollution. They have a maximum speed of 200mph. What are the benefits and problems of using airships?

5 Give the case for and against people being allowed only so many air miles of flight a year.

Will we have fish to eat?

Global demand for fish has doubled in under 30 years, because of population growth in poor countries and a matching increase in demand for fish there.

Britain will demand the right to catch more cod after the fishing industry warned that the European Union fishing **quotas** are forcing crews to dump thousands of tons of dead fish into the sea. Fishermen often catch large amounts of fish such as cod by accident when targeting other fish or shellfish and cannot land them because they have no quota or the fish are below the minimum landing size.

Britain's request differs from international scientists' advice that North Sea cod catches should be only 50 per cent of catches last year to allow a promising number of young fish to rebuild the stock. Environmentalists have branded the idea of an increased quota as 'madness', saying it would endanger the recovery of cod stocks.

Fishermen have complained about the destruction of the fishing industry and their way of life. The industries relying on fishing and local shops have lost money and as a result jobs.

6 Explain what the effect of **overfishing** will be on the sea ecosystem.

7 Explain for each case study what is meant by the phrase 'the **cost of going green**'.

8 Present a short article (150 words) on why poverty in the developing world may hinder the fight against pollution.

1 Write down each type of pollution and its correct method of control.

Type of pollution:	Method of control:
air	sewage works
water	green belt
land	catalytic converters in car exhausts

2 a Write down each cause of pollution and its effect.

Cause:	Effect:
sulphur dioxide	destruction of the ozone layer
CFCs	greenhouse effect
carbon dioxide	acid rain

b What problem occurs if the ozone layer becomes thinner and develops a hole?

c Give **two** effects of acid rain on the environment.

3 Severn Trent Water Authority samples streams and rivers. It uses the table below to judge the acidity of the ponds in its area.

Organism	Lowest pH organism can survive in
perch (fish)	4.5
brown trout (fish)	4.7
mayfly larva	5.5
water boatman	3.5
frog	5.0
fresh-water mussel	6.0

a The first river it tests has perch, trout and water boatman in. Suggest a pH value for this river.

b The second river it tests has mayfly, frogs, water boatman and perch in.

  i Suggest a pH value for the river.

  ii Which other animal species would you have expected the water authority to find? Suggest why it is not present.

c No life is found in a third river. Suggest a pH value for the river.

4 The number of plants in two different areas of meadow is measured using a quadrat and a tape measure. Sophie and Lance use the following method.

- Place the quadrat at random in the area.
- Count the number of different species in an area.
- Repeat this another nine times.

a What is a quadrat?

b Why are the results of ten quadrats taken?

c Why is the quadrat placed at random?

d How was the quadrat placed at random?

5 Over the last 30 years the pH of some lakes in southern Norway has been measured and the number of fish in the lakes has been recorded.

% Number of lakes

pH of lake	good fish population	poor fish population	no fish
4.4	3	41	56
4.7	6	38	56
5.1	16	48	36
over 5.5	71	8	21

a What pattern is shown by the results?

b Suggest why this pattern has occurred.

c How is acid rain caused?

6 Green belts are areas for wildlife around towns and cities, and cannot be built on. Give a reasoned argument to show the case for and against the use of green belts for building.

7 a Scientists noticed that soil heaps near old lead mines had little grass growing on them compared to the surrounding land. They suggest it was due to high levels of lead salts in the soil.

   Design an experiment to see if lead salts affect plant growth. Indicate in the plan:
   - What factors you will keep the same
   - How you will set up the equipment
   - How you will measure your results
   - How you will avoid anomalies.

   b The scientists found the soil was also very acid in plants, pH3. They wanted to investigate whether it was the pollution of the lead salts, the acid or the interaction of both covering the bare areas.

   Plan an investigation to find out the major cause of these bare areas. How will you determine which has the greatest effect?

Topic Summary

Learning Checklist

☆ I know **two** materials that are recycled.	page 174
☆ I know that burning fossil fuels causes pollution.	page 176
☆ I know **one** way to conserve plants and animals.	page 180
☆ I can name **one** endangered species.	page 180

☆ I know the effect of acid rain.	page 170
☆ I know how to control the production of acid gases.	page 170
☆ I know what the ozone layer does.	page 175
☆ I know the effect of global warming.	page 177
☆ I know how to sample an area.	page 179
☆ I know how endangered species are conserved.	page 180
☆ I can give **two** examples of conservation.	page 180

☆ I can explain why pollution occurs.	page 169
☆ I can explain how acid rain is caused.	page 170
☆ I can explain why global warming occurs.	page 177
☆ I can explain the purpose of conservation.	page 181

☆ I can give a reasoned argument for the need to control carbon dioxide emissions	page 177
☆ I can explain what is meant by the 'cost' of becoming green	pages 180–181
☆ I can give a reasoned argument for conservation.	page 181

☆ I can analyse evidence from different 'pressure' groups, to produce a reasoned argument for or against an environmental problem.	page 173
☆ I can understand why scientific ideas may change in relation to the short- and long-term effects of the environmental change caused by pollution.	page 183

☆ I can plan complex investigations to identify the cause of environmental problems.	page 182
☆ I can explain how scientific knowledge of global climate change has been developed through evidence gathering, investigating and questioning our present understanding.	page 183

Topic Quiz

1. Name **two** plants found in grassland.
2. Name **two** materials that are recycled by a council.
3. Why is ultraviolet light dangerous for humans?
4. Give **two** problems caused by acid rain.
5. Give **two** problems caused by global warming.
6. Name the metal grid used to sample an area.
7. How is acid rain caused?
8. What does 'protected species' mean?
9. What is the greenhouse effect?
10. Explain why global warming is occurring.
11. How can acid gas emissions be reduced?
12. What does 'conservation' mean?
13. Give **two** ways of conserving natural areas.
14. Why should the blue whale be conserved?
15. How can you find the diversity of plants in an area?

True or False?

If a statement is false then rewrite it so it is correct.

1. Fossil fuels are the main cause of acid rain.
2. The blue whale is an example of a species that is extinct.
3. A quadrat is used to count the plants found in an area.
4. Global warming is caused by the greenhouse effect.
5. Carbon monoxide causes acid rain.
6. All plastics are biodegradable.
7. Limestone can be used to neutralise acidic lakes.
8. The ozone layer blocks out heat from the Sun.
9. A car produces more pollution travelling through cities than at a steady 50 mph on a motorway.
10. The number of polar bears is falling because they are being over-hunted.

Numeracy Activity

John obtained the following results when he sampled his lawn.

Quadrat number and % cover of the plant

Plant found	1	2	3	4	5
daisy	5	3	4	1	0
clover	15	6	7	10	7
dandelion	0	8	3	0	5
grass	80	79	75	84	83
plantain	5	2	8	1	5

Work out the average percentage cover for each plant. Then draw a bar graph of your results.

ICT Activity

Use the Internet to find out the air pollution levels in your area for a period of five weeks.

Vesuvius

Vesuvius is one of the most active volcanoes in Europe. It dominates the Bay of Naples in southern Italy. Old paintings of Vesuvius show that both the size and shape of the volcano have changed completely over the centuries. The geological changes associated with this famous volcano have influenced our views about history, art and the way the Earth works.

In 1830 Charles Lyell published one of the first scientific books about geology. He noted some very strange observations at the Roman temple of Jupiter Serapis, right at the foot of Vesuvius. The temple columns were then 6 metres above sea level but the stones had mysterious holes bored in them. Theses holes contained the shells of dead sea creatures. The evidence convinced Lyell that something extraordinary had happened there. The Romans must have built the temple above sea level, two thousand years ago. After construction, the temple had sunk below the waves and later had been lifted up again, clear of the water. We now believe that these changes of ground level are associated with lava moving below the ground.

The Roman city of Pompeii as it was before the eruption of Vesuvius.

BIG IDEAS

By the end of this unit you will be able to describe the main types of rocks, explain the processes by which one sort is changed into another and combine the ideas in the Rock Cycle.

The best known evidence of the power of Vesuvius comes from the ruins of Pompeii. The eruption of AD79 was recorded by an eye-witness, Pliny the Younger. His letters describe hot ash that rained down on the people lasting for days and burying the city 7 metres deep. Pompeii is a time capsule of Roman life. It contains all the peoples' belongings and even some bodies of those unable to escape in time. Pliny's descriptions of the destruction of Pompeii match closely with modern scientific observations of volcanic eruptions in other parts of the world.

What do you know?

1   Name an active volcano in Italy.

2   How do we know that volcanoes can change in appearance?

3   Where was the Roman temple of Jupiter Serapis?

4   What observation did Charles Lyell make at the temple?

5   What conclusions were based on his observations?

6   What category of rock is formed by volcanic activity?

7   Name three materials that might be produced during a volcanic eruption.

8   What was Pliny's contribution to understanding what happened at Pompeii?

9   What would happen to an existing rock such as limestone near an active volcano?

10 What category of rock might be forming in the sea near an active volcano, as a result of weathering?

Studying sedimentary rocks

BIG IDEAS

You are learning to:
- Describe what happens to rock particles during weathering
- Provide evidence that rock particles can form new rocks
- Explain the link between the particles in a rock and the rock's properties

How hard is breaking up?

When old buildings are knocked down, brick walls are broken up to leave smaller pieces. If the bricks are valuable they may be broken off one by one for recycling. The bricks are like the **particles** making the wall. **Sedimentary rocks** are also made of particles. The particles may be sand grains if the rock is a sandstone. The particles are held together by natural cements such as calcite or silica. The chemical in calcite is the same as that in chalk or limestone (which are both examples of sedimentary rocks).

FIGURE 1: Breaking up buildings.

1 What are the small pieces that make a rock called?

2 How do these small pieces stay together?

Rocks from rivers

FIGURE 2: How do rivers cause changes to the rocks in an area?

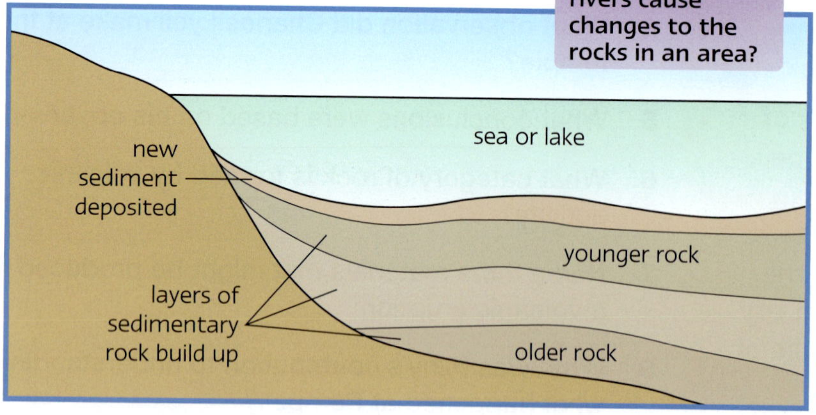

Diagram labels: new sediment deposited · sea or lake · younger rock · layers of sedimentary rock build up · older rock

How Science Works

Homemade rocks

Running water down a sloping plastic gutter is a model for a river. A clear plastic tank or bottle can collect the water, acting like the sea.

Method
- Set up a 2-metre plastic gutter on a gentle slope with a plastic tank at the lower end.
- Spread a mixture of clay, sand and stones of varying sizes along the gutter.
- Switch on the water at the top end, using a spray head like a watering can.
- Observe how the sediments behave and where they go.
- Record all your observations and try to explain them.

1 Which material was moved most easily by the model river?

2 Which material needs the most energy to move it?

3 Describe the appearance of the new sediment that collected in the tank. Comment on any layers you can see there.

HSW

Rivers are important causes of geological change. They move particles of old rocks and deposit them in the sea. The process of laying down new layers of sediment is called deposition. As the weight of the new sediment increases, the layers below are compressed. Fast flowing water can move large stones but when the water slows down, the stones sink. This is why you often find pebbles deposited first when a river meets the sea. Sediments form in layers called **strata**, the

... breccia ... cleave ... conglomerate ... fossil

oldest at the base. The particles become finer the further out to sea it is. Fine particles turn into mudstone. Pebbles can become a **conglomerate**.

As new layers collect, the older strata below are squeezed and water is expelled.

In the mountains, the rough pieces of rock may become the rock called **breccia**.

3 What name is given to layers of sedimentary rock?

4 What is the difference between a conglomerate and a breccia?

5 Why do mudstones form out at sea?

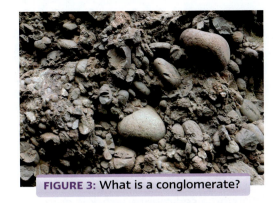
FIGURE 3: What is a conglomerate?

Splitting up

The force of gravity means that sedimentary rocks are deposited in horizontal layers. Sometimes earth movements raise these strata above sea level, as at Lyme Regis. The rocks there are marine limestones and shales, a rock with fine particles. If you use a chisel you can **cleave** (split into layers) the rocks. Often at Lyme Regis **fossils** are found inside.

Sedimentary rocks often contain characteristic fossils. The fossils give us information about the conditions on Earth when they lived.

6 In terms of particles, what happens when you cleave a rock like slate?

7 What can you deduce from seeing vertical sedimentary strata in a cliff?

8 What ecological information might you obtain from a study of fossils?

FIGURE 4: What are these rocks at Lyme Regis famous for?

FIGURE 5: An ammonite fossil.

The bigger picture

Rocks that form cliffs and hills are likely to be resistant to erosion. If the rocks can be cut into regular shaped blocks, they will be suitable as building stones. The limestone known as Portland Stone is a famous example. It forms a peninsula of resistant rock, almost surrounded by the sea. Portland Stone has been used to create many famous buildings such as St Paul's Cathedral in London. When rocks can be quarried near the sea, as in Scandinavia, it becomes economic to import them. Scandinavian rocks are used in Britain to make sea defences where local rocks are unsuitable.

9 How does the following general reaction explain what happens to Portland Stone in industrial cities?
carbonate + acid ➔ water + carbon dioxide + a salt

How Science Works

Most rock strata you see in cliffs are at an angle, not horizontal. As you walk along a beach, you keep seeing rocks of different ages, one under the other. Geologists study these strata and build up a complete history of the area, called the succession. The succession also shows how climate changes over millions of years.

... particle ... sedimentary rock ... strata

More about sediments

BIG IDEAS

You are learning to:
- Identify a sedimentary rock
- Interpret the messages that rocks can tell us
- Recognise that rocks store clues about the climate

Rock watching

Spotting a sedimentary rock is quite easy. These are the clues to look for:

If you can answer 'yes' to the questions, it is a sedimentary rock such as **sandstone**.

> Is it made of particles?
> Can you see layers?
> Are there any fossils?

Granite is different – it is made of **crystals** and there are no layers and no fossils.

FIGURE 1: Which sample is sandstone?

1 Would you find fossils in granite?

2 Give **two** properties of sedimentary rocks.

Fossil information

When sedimentary rocks form they often contain **trace fossils**. These give clues about the environment of formation. Some examples include mudcracks preserved in rocks, rainprints, impressions of leaves and even sets of footprints. A set of dinosaur footprints can tell us quite a lot!

The size of a dinosaur can be estimated from its footprints and its weight can be estimated from the depth of the footprints. There may even be evidence of a tail dragging along the ground.

Fossils can tell us if the rock formed on land or under the sea. Colours are useful too. Red sandstones form in deserts, such as the Sahara. Coal is a useful sedimentary rock formed from plant remains in swamps. Coal miners often find tree roots, fossil wood and leaf prints in the coal.

How Science Works

Rock watching

You will need a selection of unlabelled rock samples that you can examine using a hand lens or magnifying glass. Suitable rocks include sandstone, limestone, shale, breccia and conglomerate.

Method
- Note the colour and if there are any particles.
- Test the hardness using the point of an iron nail to see if the rock can be scratched.
- Look for evidence that the particles are in layers.
- Check for any fossils or fossil impressions.
- Tabulate your findings and say which of your samples are likely to be sedimentary rocks.
- Why might footprints in wet concrete puzzle people studying rocks in the future?

FIGURE 2: Mudcracks can give clues to geologists.

FIGURE 3: Fossilised dinosaur footprints – what can they tell us?

... crystal ... fossil record ... granite

3. What kind of fossil is a set of footprints?

4. Devon cliffs are made of red sandstone. What does this tell us about the climate when they formed?

5. How do we know coal is a sedimentary rock?

What about us?

Certain parts of the world are rich in fossils of one kind. In China large numbers of dinosaurs and clutches of dinosaur eggs have been found. In East Africa fossils related to our own species have been found. Fossilisation is a rare event, most plants and animals simply rot away leaving no trace. The **fossil record** will always be incomplete.

Recent discoveries near Lake Turkana in Africa suggest that two different kinds of early human-like creatures lived at the same time. It may just be luck that our own ancestors survived.

6. What special fossils have been found in China?

7. Why is the fossil record incomplete?

8. Why are conclusions about evolution only estimates when they are based on the fossil record?

Organic clues

Crude oil and natural gas are usually found in sedimentary rocks. It is difficult to find out what is contained in rocks that are many kilometres underground. Some sedimentary rocks give us clues to follow up. On the Dorset coast there are oil seepages from rocks along the shore. Examining the rocks shows that they contain trapped droplets of oil. This gave geologists the idea that there might be an oilfield underground. The same rocks extend under Poole Harbour and there is a major oilfield currently in production there. All of the machinery to extract the oil is hidden away, surrounded by woods.

FIGURE 4: Fossil of an early human-like ancestor.

9. How might oil be able to move through a sedimentary rock?

10. What other interpretation could be made from finding oil on the beach?

11. What is the meaning of the term 'reservoir rocks' and what is their importance?

... sandstone ... trace fossil

People have always wondered about the age of the Earth. Some religions suggest that it is just a few thousand years old. At the end of the nineteenth Century, physicists calculated how long it had taken for the Earth to cool to its present temperature. They assumed that it started as a red-hot ball of molten material. The calculation gave an age of several million years. Unfortunately this was before the discovery of radioactivity. All rocks are slightly radioactive and when radioactive atoms decay, they release heat. When this was included in the calculation it put back the Earth's age to several thousand million years. All of these figures are estimates based on scientific observations and some assumptions. One assumption is that the laws of physics have always been the same as they are now.

There are further problems explaining how coal deposits could form in Antarctica or in Spitzbergen near the Arctic Circle. Theories have suggested that the Earth's climate was once much hotter or that the continents themselves have moved around the globe. Surprisingly, the evidence for continental drift is very strong. Look at the shapes of East Africa and South America in an atlas, they could fit together like a jigsaw. This theory was suggested by Alfred Wegener in 1915 but was dismissed at the time as being impossible.

Assess Yourself

1 According to religious views, could people have lived alongside the dinosaurs?

2 How did Victorian physicists estimate the age of the Earth?

3 Which discovery undermined the Victorian calculation?

4 What important assumption do we make about the laws of science when considering long periods of time?

5 Coal forms in tropical swamps. Why is this a problem for geologists?

6 What did Wegener notice about the continents?

7 Why do you think that Wegener's theory was first rejected but later accepted?

8 What evidence about the age of rocks might you obtain by observing how fast sediment from a river collects on the seabed?

History Activity

Research the theory of continental drift as first proposed by Alfred Wegener. Concentrate on identifying some evidence that he put forward in support of his ideas.

Geography Activity

Study a map of the continents to see how well Africa and South America might have fitted together. Cut out copies of the continents to check. There is a better fit if your map shows the edges of the continental shelves.

Level Booster

EP Your answers show that you can critically compare competing theories and justify your own views.

8 Your answers show an understanding of the provisional nature of scientific judgements and that new evidence may overthrow them.

7 Your answers show an advanced understanding of the complex ideas involved in dating the Earth.

6 Your answers show a good understanding of the importance of evidence in supporting scientific theories.

5 Your answers show a good understanding of the competing theories of Earth history.

4 Your answers show a basic understanding of theories of Earth history.

Mountains and folds

BIG IDEAS

You are learning to:
- Use a model to understand how folds might form
- Discuss the nature of folding
- Evaluate the evidence for continental drift

Blanket battles

Most people arrange clean blankets or towels folded into neat piles. All the blankets start flat, forming one layer each. Sedimentary rocks look a bit like this. When you push the blankets in from the edges, they squeeze up into folds. Mountains like the Alps and the Rockies are called **fold mountains**. Earth movements have squeezed the layers of rock into gigantic folds, the size of mountains. The Alps are so old that the top halves of the folds have been worn away by the weather.

1 Why are piles of blankets like rocks?

2 What happens to rock layers when they are squeezed?

Folded rocks

Folding can also be seen on a small scale in coastal cliffs.

- When the **strata** (layers of rock) are pushed up into a dome shape, we call it an upfold or **anticline**.
- When strata are forced down into the shape of a bowl, it is called a **syncline**.

FIGURE 1: Folding in the cliffs at Lulworth Cove in Dorset.

FIGURE 2: The Himalaya range in Nepal has the highest mountain on Earth – Everest.

If you fly over mountains you can often see chains of fold mountains. The highest mountains on Earth in the Himalaya range are examples of fold mountains.

Sediments and lava flows are generally deposited in horizontal layers owing to the influence of gravity. Earth movements cause them to bend, tilt or even fracture into pieces. These are **tectonic** structures, linked to earth movements.

... anticline ... fault ... fold mountain ... folding

When rock layers fracture, a **fault** may be formed. On one side the rock layers may slip down, shown by a break in the continuous strata.

3 What is an upfold in rocks called?

4 What type of mountains are the Himalayas?

5 How could you identify a fault in a cliff?

Squeezed dry

When new loose sediments are deposited by rivers in the surrounding oceans, water is trapped too. As new layers form on top, the enormous pressure squeezes the lower layers. The rock particles get closer together, water is pushed out and natural cements such as calcite fix the rock particles in position. The greater the weight of rock layers above, the more the lower strata are compressed. We sometimes see fossil shells that have been squashed completely flat by this process.

FIGURE 3: What causes a fault to appear in layers of rock?

6 Why does the water content of new sediments change?

7 How might some fossils in lower strata be damaged or deformed?

Drifting along

Convection currents within the Earth's mantle, below the crust, are strong enough to move the continents. The movements are slow, a few centimetres each year. The Earth's surface is divided into enormous sections called tectonic plates. As they slide along or crash into each other, rocks are crumpled, folded and compressed. Rocks and fossils in Africa and in South America match each other exactly. This suggests that these land masses have drifted apart, having once been joined together in a single continent.

8 What other evidence did Wegener put forward to support the idea of continental drift?

9 Why do oil companies look for anticlines when hunting for new oil deposits?

10 Why is it difficult for palaeontologists (studying fossils) to understand the 3-D shapes of fossil creatures?

... strata ... syncline ... tectonic

Metamorphic rocks

BIG IDEAS

You are learning to:
- Describe how rocks are changed
- Recognise the features of metamorphic rocks
- Make connections between the properties and uses of metamorphic rocks

All change

When clay is wet you can shape it easily. It is easy to squash it down and start again. Once the clay pot is ready it is put in an oven. The heat changes the clay completely. After heating, the clay is hard and **brittle** – it breaks if you drop it. **Heating** changes the material. Rocks can also be changed by heating. Changed rocks have a special name – they are **metamorphic** rocks.

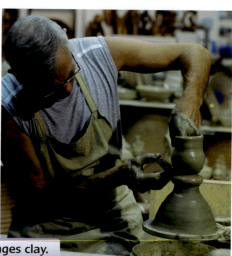

FIGURE 1: Heat changes clay.

1 Give **one** difference between wet clay and the same clay after heating.

2 What are metamorphic rocks?

Original material	After metamorphism	Uses of the metamorphic rock
Sandstones, sand grains in layers	Quartzite, much harder, original layers destroyed	Building stone
Limestone, layers (strata), often with fossils	Marble, much harder, shiny, no fossils left	Building stone, statues, work surfaces
Mudstone, layers (strata), soft, crumbles easily	Slate, very hard, shiny, splits in a single direction to give flat sheets	Roofing, facings for buildings

Metamorphic properties

When existing rocks are heated, for example by hot molten rock nearby, they **recrystallise** and new crystals form. This is what happens around volcanoes. The structure of the original rock changes permanently. Metamorphic rocks are usually very hard and shiny. The minerals that form crystals in the new metamorphic rock give clues about its origin. Different minerals form at different temperatures. There are often bands of minerals in metamorphic rocks which change the further away you get from the heat source.

Limestone, chalk and marble are chemically identical but marble has been metamorphosed. It can be polished and is extremely hard. Marble has been chosen by sculptors because it can be carved into complex shapes.

Did You Know…?

The heat and pressure of metamorphism can produce gemstones for jewellers. The deep-red gem called garnet was very popular in Victorian times. Metamorphism causes rocks to form new crystals, to recrystallise, just like the garnets.

… brittle … gneiss … heating

Metamorphism can involve both heat and pressure. This is how soft mudstone turns into hard slate for roofing. Slate can be split easily into thin sheets – it cleaves easily.

3 Why are volcanoes and metamorphism sometimes linked?

4 Give **two** properties that may change when a rock is metamorphosed.

5 Give **one** similarity and **one** difference between chalk and marble.

FIGURE 2: Shiny metamorphic rock. How is metamorphic rock formed?

FIGURE 3: Marble is used for work surfaces. Why is it suitable?

FIGURE 4: Slate is a metamorphic rock. How does slate form?

Metamorphic information store

Metamorphism can vary a great deal, there are many possible permutations of temperature and pressure. Each set of conditions produces different rocks. The most intense metamorphism is called high-grade and produces **gneiss**. Gneiss (pronounced 'nice') has alternating bands of light and dark minerals. This intense metamorphism is associated with the formation of new mountains.

Metamorphism can destroy information contained in rocks. Limestones that are full of marine fossils may metamorphose into marbles that are fossil free. The heat and sometimes pressure are enough to destroy traces of fossils and hence clues to the origin of the rock.

6 What does gneiss look like?

7 Why are there so many different kinds of metamorphic rocks?

8 Marble forms from limestone but where are the fossils?

How Science Works

Examine samples of all the rocks described in the table. Note the colours, hardness and any evidence of fossils. Tabulate your results.

HSW

FIGURE 5: Gneiss rock.

Rock detectives

Rocks contain lots of evidence of their past histories. The heat from volcanoes can form a range of new minerals in existing rocks. The directions in which slate can be cleaved gives evidence of the tectonic forces that changed the original mudstone. Distorted traces of fossils may show the effects of intense heat and pressure during metamorphism.

9 **a** There are some famous marble quarries in northern Italy. Suggest a possible geological history for this area.
b Describe three characteristics of slate that make it a suitable roofing material.

10 A geologist found a rock called hornfels that is only produced by intense heat. What might have formed the hornfels?

Crystals in igneous rocks

BIG IDEAS

You are learning to:
- Explain why crystals vary in size
- Compare different types of crystal
- Explain why crystals have the same basic structure

Crystals in all sizes

There is something mysterious about the sizes of crystals in **igneous** rocks. When hot molten rock, as in volcanoes, cools down, crystals are formed. The rate of cooling determines the crystal sizes. Below ground, rocks can stay hot for thousands of years. At the surface, lavas may cool in hours or weeks giving little time to grow large crystals.

- **Granite** cools slowly deep underground giving plenty of time for large crystals to grow.
- Lava thrown out of a volcano cools rapidly. There is no time for crystals to form before it sets solid. **Obsidian** is a natural volcanic glass and **basalt** is the most common type of lava rock. Neither shows many crystals at all.

1 What is the most common type of lava rock?

2 What is the most obvious difference between granite and obsidian?

Quartz everywhere

The overall shape of a crystal depends on the relative sizes and shapes of the particles it contains. The particles pack closely together and form regular 3-D patterns. The shapes of the crystals of different minerals formed in rocks can vary considerably.

3 When thick glass breaks it gives a characteristic curved edge. Why does obsidian break in the same way?

FIGURE 2: Quartz crystals.

4 What would happen to the crystals if you melted basalt and let it cool very slowly?

5 Why are the crystals formed by cooling magma that is underground usually different from identical magma when it escapes as volcanic lava?

How Science Works

Use the particle model to explain crystal shapes.

FIGURE 1: Which is granite and which is basalt? (Hint: remember granite has large crystals.)

Did You Know...?

Mica is one of the three minerals making up granite. When mica crystals grow very large they are mined and have many uses.

- Muscovite mica is clear and was used for the windows in solid fuel stoves since it withstands high temperatures.
- Mica is a good electrical insulator and scrap mica is made into artificial snow for Christmas trees.

Grow your own crystals

You can grow instant crystals and investigate how temperature affects the crystal size.

This is like the formation of igneous rocks in nature.

Method:

1   Put five spatulas of salol crystals into the test tube.
2   Put the test tube into hot water (70 °C) to melt the crystals.
3   Using a dropper, put a large drop of salol on each slide.
4   Observe how the crystals form on the cold and the hot slides.
5   If crystals are slow to form, add one seed crystal of the original salol.
6   Record your observations.

Questions

1   How do the patterns of particles inside large and small salol crystals compare?

2   Explain why hot and cold slides produce salol crystals of different sizes.

3   Which temperature slide most closely resembles what happens with lavas?

Particle pictures

When you look closely at crystals you notice some key features. Crystals of the same chemical may vary in size but are all the same shape or crystal habit. The angles between flat crystal faces are also constant for a particular mineral. An unlikely experiment gives a clue about these shapes. If you pour lots of table-tennis balls (the particles) from a bucket into a shallow dish, you notice something odd. The particles just don't go anywhere. They start to build up into a 3-D pattern, a large model of a crystal. On this scale, a salt grain would be about the size of the Moon. Real crystals owe their shapes to the regular 3-D arrangement of the particles inside.

Crystal growth

A cross-section through a lava flow often shows a variation in crystal size. Where the lava cooled quickly, near the edges, crystals are small or absent. In the middle, where heat was lost more slowly, crystals had time to start growing.

6 Volcanic explosions can produce volcanic bombs, bits of molten lava thrown up into the air. Why do they often have a glassy texture?

Volcanic magic

BIG IDEAS

You are learning to:
- Describe the inside of a volcano
- Explain how volcanic action forms rocks
- Make connections between mineral formation and temperature

Inside a volcano

The inside of a volcano is a dangerous place. There are made-up stories about reaching the centre of the Earth through a volcano. The heat and fumes would kill off any travellers who tried. Volcanoes are made of layers of **ash** and **lava**. Every time the volcano **erupts**, a new layer is added on top. Poisonous gases and ash and even volcanic bombs (lumps of hot rock) may all be thrown out of a volcano. It was ash that destroyed the Roman city of Pompeii, near Vesuvius.

1 Why is it unlikely that explorers can enter an active volcano?

2 What might have killed the people of Pompeii?

New rocks

Iceland is on the junction between two tectonic plates. As these plates move apart, basalt lava pours out from the magma chamber to fill the gap. Plate boundaries are often associated with active volcanoes, hot springs, geysers and also with earthquakes. The heat of the rocks causes the metamorphism of any existing rocks in the same area.

3 Give three features commonly found near plate boundaries.

4 Where do we find magma chambers?

5 Why is the size of Iceland likely to grow in the future?

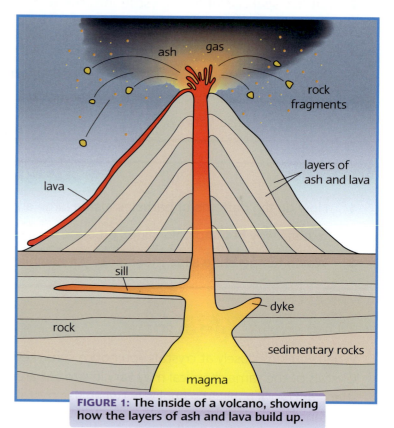

FIGURE 1: The inside of a volcano, showing how the layers of ash and lava build up.

FIGURE 2: Surtsey – a brand new volcanic island.

... ash ... erupt ... fluid

Cornish miners

Magma often stays buried underground and cools slowly. As it cools, **fluids** escape into surrounding rocks, both gases and liquids. These hydrothermal fluids can be hot enough to cause local metamorphic changes to the surrounding rocks. In addition, the fluids can have important economic consequences. The fluids can deposit new minerals containing tin, copper, lead, arsenic and even radioactive uranium. Tin was mined in Cornwall for thousands of years. Rings of similar minerals are deposited as the spreading hot fluids gradually cool down. Finding tin ore, called cassiterite, in one area can give clues as to where to look in other places where mineralisation occurred.

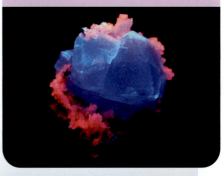

FIGURE 3: A disused tin mine in Cornwall. Where did the tin come from?

6 What is produced when magma reaches ground level?

7 What category of rock will be formed by contact with a magma chamber?

8 What is the economic importance of fluids from igneous rocks?

Seeing a pattern in volcanism

Mapping the locations of active volcanoes worldwide and the occurrence of earthquakes shows a very clear pattern. Volcanoes are common on plate boundaries. As the plates shift, often violently, earthquakes are produced. Earthquakes under the sea can result in a tsunami, a destructive giant wave.

9 What deductions might you make from the location of an extinct volcano?

10 How might the pattern of mineralisation in Cornwall provide evidence for the existence of hydrothermal fluids?

The rock cycle

BIG IDEAS

You are learning to:
- Explain that rocks can be 'secondhand'
- Discuss how the rock cycle works
- Use the idea of the rock cycle to explain how rocks can change

The rock cycle

The rock cycle is a useful way to picture the relationships between rock types. The first rocks to form are always igneous rocks, as in a volcano. The rapid cooling of these extrusive igneous rocks such as lava, gives glassy rocks with few or very small crystals. The particles in the lava cool too rapidly for large crystals to grow. Underground, the insulating effect of the rocks above produces slow cooling. Intrusive igneous rocks such as granite therefore have large crystals.

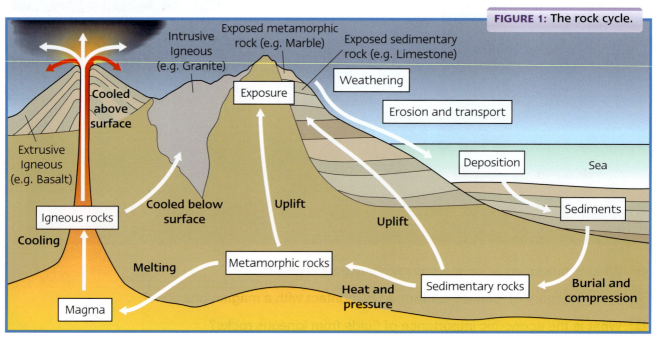

FIGURE 1: The rock cycle.

Rocks at the surface are weathered and rock particles are transported by rivers and deposited as new strata of sedimentary rocks.

1. Why do the thicknesses of new sediments change with time as new strata form above them?

2. At what stage might fossils be destroyed in the rock cycle?

3. Where on a tectonic plate are you most likely to find volcanoes?

... crust ... extrusive

Rocks and their ages

In cliffs made of sedimentary rocks, the oldest rocks are generally at the base since they formed first. Earth movements can turn strata upside down or move them from horizontal to vertical. The forces involved in the movement of tectonic plates can turn old strata into brand new mountains. Both tectonic movements and the heating effects of the upper mantle can melt rocks. The nature of the minerals formed gives evidence of the temperatures and the pressures when new rocks were being formed. Metamorphism can occur more than once in the same area. This means that metamorphic rocks preserve only the evidence of the most recent metamorphic event.

4 Which parts of the rock cycle are hardest to investigate near an active volcano?

5 What evidence about the age of the rocks in an area might be gained by studying the rate of growth of new sediments?

6 What are the limitations of modelling the way igneous rocks from by studying the behaviour of molten salol?

7 Why do crystals of the same mineral always have the same shape but are often different sizes?

8 Describe two possible new rocks that could form from an existing limestone.

Lunar geology

HSW

FIGURE 2: Limestone

1 For each of the following statements, write 'T' if the statement is true or 'F' if it is false.

a Sedimentary rocks form in layers.

b Marble forms from limestone.

c Limestone is a metamorphic rock.

d Slate splits easily into sheets.

2 Use the following words to help you answer the questions. The words may be used once, more than once or not at all.

anticlines   fault   folded   fossils   steam

a Limestone contains the remains of living things called _____ .

b Layers of rock can be _____ by earth movements.

c A break in rock layers is called a _____ .

d Ash, _____ and lava are produced by volcanoes.

3 Write down each rock or mineral and choose its correct description.

basalt	sedimentary rock
slate	mineral
conglomerate	igneous rock
quartz	metamorphic rock

4 Write down each rock or mineral and choose its correct use.

limestone	artificial snow for Christmas trees
quartz	building sand
mica	to build cathedrals
slate	for roofing

5 Salol crystals are melted in a hot water bath. Drops of salol are added to cold or hot glass slides. Write 'T' for any statements that are true and 'F' for any that are false.

   a Crystals form more quickly on the cold slide.

   b This is a model to represent the formation of metamorphic rocks.

   c Crystals form more slowly on the hot slide.

   d Salol melts above 100 °C.

6 Fill in the missing words for the following statements about the rock cycle.

   a The rock cycle begins with _____ rocks.

   b Sandstone is one example of a _____ rock.

   c During metamorphism, sandstone turns into _____ .

   d The _____ chamber supplies molten rock to a volcano.

7 This question concerns theories about the Earth.

   a How could you check Wegener's theory by using an atlas?

   b How old did the late nineteenth Century physicists think that the Earth was?

   c What factor had the physicists failed to take into account?

   d How can you account for this mistake?

8 **a** Why is the cleavage of slate a useful property?

   b How does pottery provide an analogy for metamorphism?

   c What evidence is there of rock processes similar to Earth on other planets?

   d Why was tin found in Cornwall?

9 This question concerns lunar geology.

   a Why is it easier to interpret ancient surface rock features on the Moon?

   b What evidence would you look for to see if there was ever flowing water on the Moon?

   c What use might a lunar space station make of the location of a volcano?

   d What would be the significance of a lunar fossil?

Learning Checklist

4

☆ I know that there are three rock types.

☆ I know that some rocks form in layers.

☆ I know that some rocks can melt.

☆ I know that heat changes the type of rock.

☆ I know that there is a rock cycle.

5

☆ I know that there are igneous, sedimentary and metamorphic rocks.

☆ I know that sedimentary rocks form in layers or strata.

☆ I know that there are different theories of the age of the Earth.

☆ I know that metamorphism produces new rocks.

☆ I know that metamorphic rocks are often hard and shiny.

☆ I know that igneous rocks were once molten.

☆ I can describe the stages of the rock cycle.

6

☆ I know the characteristics of the three rock types.

☆ I know that sedimentary rocks can contain fossils.

☆ I can describe some theories to explain the age of the Earth.

☆ I can explain how material is recycled in the rock cycle.

7

☆ I can identify the three rock types from descriptions.

☆ I know why only sedimentary rocks have fossils.

☆ I understand the uncertainties surrounding the age of the Earth.

☆ I know that metamorphic rocks have uses related to their properties.

☆ I can distinguish between types of igneous rocks.

8

☆ I can argue for a particular theory of the Earth, using a range of evidence.

☆ I can explain crystal habit in terms of particles.

☆ I can relate the locations of metamorphic minerals to the source of the heat.

Topic Quiz

1. Name **one** volcano in Europe.
2. Which Roman city was destroyed by a volcano?
3. Why was the city preserved for so long?
4. Name **three** materials produced by volcanoes.
5. Why are sedimentary rocks produced in horizontal layers?
6. What information would a fossil give you about a rock?
7. What kind of mountains are the Alps and the Himalayas?
8. What is a break in layers of strata called?
9. Why do igneous rocks have crystals of varying sizes?
10. What size crystals would you find in an extrusive igneous rock?
11. Which type of rock begins the rock cycle?
12. What did Wegener notice about the continents?

True or False?

If a statement is incorrect then rewrite it so it is correct.

1. Granite is an igneous rock.
2. Slate is a sedimentary rock.
3. Volcanoes produce igneous rocks.
4. Fossils help us date a rock.
5. There are three different minerals in granite.
6. Intrusive igneous rocks often have small crystals.
7. Metamorphic rocks can melt to give igneous rocks.
8. Upfolds are also called anticlines.
9. Wegener's theory of continental drift was accepted by his contemporaries.
10. Volcanoes can turn limestone into marble.
11. Crystals in volcanic sills are usually small.

Literacy Activity

Describe the experience of discovering a new dinosaur fossil that will be named after you.

ICT Activity

Carry out a web search on the theory of Alfred Wegener (1915) about drifting continents.

Present your findings as a PowerPoint presentation.

Glossary

Keyword	Definition	Page
Acid rain	Rain that has a pH of less than 7	p.168, 174
Alkali	A substance with a pH of more than 7	p.90
Alkali metal	A metal which burns to form an alkaline oxide, e.g. potassium	p.62
Alloy	A mixture of metals	p.78
Anodise	To increase a metal's resistance to corrosion	p.80
Artificial selection	Producing offspring with the desired characteristics by choosing the parents	p.54
Asexual reproduction	Reproduction without sex, involving one parent e.g. taking cuttings	p.52
Autism	A rare disorder characterised by failure to develop normally in language or social behaviour	p.34
Base	An alkali	p.91
Biodiversity	A term used to describe the range of different organisms in an area	p.176
Blood Alcohol Concentration	A measurement (usually taken as mass per volume) of how much alcohol is concentrated in the blood – it is indicative of the effect of alcohol on the body	p.10
Brain	The organ that controls the rest of the body	p.22, 24, 32
Breed	To produce offspring	p.54, 178
Bronchitis	Inflammation of the air passages of the lungs (bronchi)	p.12
Cancer	A dangerous illness – cancer cells grow out of control to make lumps of cells and can invade normal healthy areas of the body	p.12
Carbon footprint	The total amount of greenhouse gases emitted as a direct or indirect result of an individual or group	p.158
Central nervous system	The brain and spinal cord	p.22

Glossary

Emphysema	Lung condition causing breathlessness	p.12
Energy	Something that does "work", including causing change such as heat or light	p.152
Evaporation	When a liquid turns into a gas without boiling	p.102
Environment	Everything a living thing is exposed to	p.42, 46
Environmental	Relating to the environment	p.42, 166
Fault	Fracture in the Earth's crust	p.194
Fertilise	What a sperm does when it joins with an egg	p.46, 48
Fibreoptics	Relating to the transmission of light and images through glass or plastic fibres	p.134
Filtration	Separation of a solid from a liquid	p.97, 102
Fold	Bends in layers of rock caused by pressure within the Earth's crust	p.194
Force	A push, pull or twist which is able to change the velocity or shape of a body	p.110, 119, 123, 154
Fossil	The preserved remains of a living thing	p.188, 190, 191
Fulcrum	The point of support (pivot, or lever)	p.120
Galvanise	To render iron rust-proof	p.80, 81
Gene	A section of DNA that controls an inherited feature	p.42, 48
Genetic	Inherited	p.43, 46
Global warming	The gradual temperature rise in the Earth's atmosphere and	p.174
Greenhouse gases	Greenhouse gases such as carbon dioxide trap solar energy an oceans create the greenhouse effect	p.175
Hydraulic	Worked by changing the pressure in a liquid	p.114
Hydroxide	A chemical containing an 'OH' group; often alkaline	p.62
Insoluble	Does not dissolve	p.100
Identical twins	Twins developed from a single egg and sharing the same genetic information	p.47
Igneous	Formed from magma	p.198
Inbreeding	Breeding with close relatives	p.55
Incomplete dominance	In genetics, a gene that does not exert total dominance e.g. gene controlling brown eyes	p.49
Infrared	A type of electromagnetic radiation given off by hot objects	p.156
Innate	A behaviour you are born with	p.26
Isobar	Lines connecting places with the same atmospheric pressure	p.112
Kinaesthetic	Learning by doing, e.g. doing experiments	p.35

Glossary

Reaction	Interaction between two or more atoms, ions, or molecules, resulting in a chemical change and the formation of a new substance	p.65
Reactivity trend	Series produced by arranging metals in order of their ease of reaction with oxygen etc	p.63, 69, 77
Recessive	In genetics, a gene that does not show when you inherit it, unless its partner gene is also recessive	p.49
Reflex	An automatic reaction	p.11, 27
Reflect	To bounce something back	p.132, 135, 145, 147, 149
Refraction	A change in the direction of a light beam, caused when it enters a material of a different density	p.141, 142
Resistance	The amount by which a conductor prevents the flow of current	p.153, 155
Second-hand smoking	Passive smoking (inhaling smoke without having smoked yourself)	p.15
Sedimentary	A new rock formed by compressing small fragments of rock like sandstone	p.189
Selective breeding	Two individuals, chosen because of their desirable characteristics, are mated together to produce offspring, hopefully with a combination of those desired characteristics	p.55
Sense organ	An organ that detects change, e.g. eyes and ears	p.23
Sex cell	Produced by the sex organs and containing chromosomes, the female sex cell is the egg and the male sex cell is the sperm	p.49
Sexual reproduction	Reproduction involving sex cells from two parents	p.53
Sill	Sheet of igneous rock created when magma intrudes between pre-existing layers of rock	p.201
Soluble	Able to dissolve	p.101, 103
Solution	Something formed when a solute dissolves in a liquid	p.65
Solvents	Liquids in which other substances can dissolve. Can also refer to a group of substances (such as glue) used as recreational drugs	p.19
Species	Group of organisms that can interbreed and produce fertile offspring	p.176, 177, 178
Stimulus	A change your sense organs can detect	p.23, 27, 31, 33
Strata	Layers or beds of sedimentary rocks	p.189, p.195
Stimulant	A drug that makes the body work faster and the heart rate increase e.g. caffeine	p.8

Acknowledgements

The Publishers gratefully acknowledge the following for permission to reproduce copyright material. Whilst every effort has been made to trace the copyright holders, in cases where this has been unsuccessful or if any have inadvertently been overlooked, the Publishers will be pleased to make the necessary arrangements at the first opportunity.

Cover photograph from NASA / Science Photo Library

The Publishers would like to thank the following for permission to reproduce photographs:

p. 6 © iStockphoto.com / Lise Gagne, © iStockphoto.com / Dieter Spears, © iStockphoto.com / Ljupco Smokovski, © iStockphoto.com / Tschon, © iStockphoto.com / Tracy Hebden; p. 8 © iStockphoto.com / Champion Photo LLC, © iStockphoto.com / Sergei Sverdelov; p. 9 © A. Dex, Publiphoto Diffusion / Science Photo Library; p. 10 © iStockphoto.com / Dieter Spears; p. 12 © iStockphoto.com / Omer Sukru Goksu, © iStockphoto.com / James Benet; p. 13 © iStockphoto.com / Nancy Louie; p. 14 © Pablo Sanchez / Reuters / Corbis; p. 15 © iStockphoto.com / maureenpr; p. 16 © iStockphoto.com / David Kneafsey, © iStockphoto.com / Tatyana Ogryzko; p. 18 © iStockphoto.com / Mariano Heluani, © iStockphoto.com / Christoph Ermel, © iStockphoto.com / Tim Starkey; p. 20 © iStockphoto.com / Hshen Lim; p. 24 © Vasily Smirnov – Fotolia.com, © Mehau Kulyk / Science Photo Library; p. 25 © James Holmes / Janssen Pharmaceutical Ltd / Science Photo Library; p. 26 © iStockphoto.com / Kenneth McAnally, © iStockphoto.com / Dzimitry Valiushka, © iStockphoto.com / TommL; p. 28 © iStockphoto.com / Viorika Prikhodko, © thislife pictures / Alamy, © Ted Kerasote / Science Photo Library, © Nina Leen / Stringer / Time & Life Pictures / Getty Images; p. 29 © iStockphoto.com / André Weyer, © iStockphoto.com / William Mahar; p. 30 © Art Wolfe / Science Photo Library, © E.R. Degginger / Science Photo Library, © Gary Vestal / The Image Bank / Getty Images, © iStockphoto.com / Karel Broz; p. 31 © iStockphoto.com / Alexander Hafemann; p. 32 © iStockphoto.com / Maartje van Caspel, © iStockphoto.com / Chris Schmidt; p. 33 © Sovereign, ISM / Science Photo Library; p. 34 © iStockphoto.com / Joshua Blake; p. 40 © Bikash Karki / WPN, © Bettmann / Corbis; p. 41 © iStockphoto.com / Brian Asmussen; p. 42 © Ian Hooton / Science Photo Library, © Alan Carey / Science Photo Library; p. 44 © Alibi Productions / Alamy; p. 45 © Gustoimages / Science Photo Library; p. 46 © iStockphoto.com / Michael Blackburn; p. 50 © Jon Freeman / Rex Features; p. 52 © iStockphoto.com / Ling Xia, © iStockphoto.com / Mark Sauerwein, © iStockphoto.com / Agita Leimane, © Nigel Cattlin / Alamy, © iStockphoto.com / Steve McWilliam; p. 53 © Najlah Feanny / Corbis Saba; p. 54 © iStockphoto.com / Eric Isselée; p. 55 © L.Beel – The Kennel Club Picture Library, © iStockphoto.com / Eric Isselée; p. 56 © iStockphoto.com / Eric Isselée, © iStockphoto.com / Scott Leigh; p. 59 © iStockphoto.com / Deborah Cheramie; p. 60 © The Trustees of the British Museum, © The Trustees of the British Museum, © The Trustees of the British Museum; p. 61 © The Trustees of the British Museum; p. 62 © Sheila Terry / Science Photo Library; p. 63 © sciencephotos / Alamy, © Stephen Finn. Image from BigStockPhoto.com; p. 64 © Martyn F. Chillmaid / Science Photo Library; p. 66 © Mehau Kulyk / Science Photo Library, © Adrian Fortune. Image from BigStockPhoto.com, © Andrew Lambert Photography / Science Photo Library; p. 67 © Magrath Photography / Science Photo Library; p. 68 © Bernard BAILLY – Fotolia.com, © Nathan Benn / Alamy, © Michael Zysman. Image from BigStockPhoto.com; p. 69 © iStockphoto.com / Jasmin Awad; p. 70 © Robert Harding Picture Library Ltd / Alamy, © iStockphoto.com / MH; p. 71 © Christophe Villedieu. Image from BigStockphoto.com; p. 72 © iStockphoto.com / Ilker Canikligil, © Andrew Lambert Photography / Science Photo Library; p. 73 © Stephen Finn. Image from BigStockPhoto.com, © Andrew Lambert Photography / Science Photo Library; p. 74 © Reuters / Corbis; p. 76 © sciencephotos / Alamy; p. 78 © 2008 Jupiterimages Corporation, © iStockphoto.com / Vadim Ponomarenko, © Nathan Benn / Alamy, © Wai Heng Chow. Image from BigStockPhoto.com; p. 79 © Dr P. Marazzi / Science Photo Library, © iStockphoto.com / Joel Blit; p. 80 © iStockphoto.com / Rob Broek, © Alistair Dick – Fotolia.com, © Franck Boston. Image from BigStockPhoto.com; p. 81 © Judy Lyon. Image from BigStockPhoto.com, © iStockphoto.com / Nicola Stratford; p. 86 © Veronique Leplat / Science Photo Library, © Robert Harding Picture Library Ltd / Alamy; p. 88 © Mehau Kulyk / Science Photo Library, © Dirk Wiersma / Science Photo Library; p. 89 © Andrew Lambert Photography / Science Photo Library; p. 90 © Gustoimages / Science Photo Library; p. 91 © iStockphoto.com / Eliza Snow; p. 92 © David Taylor / Science Photo Library, © Andrew Lambert Photography / Science Photo Library; p. 94 © iStockphoto.com / pkruger, © Eye of Science / Science Photo Library; p. 95 © Arnold Fisher / Science Photo Library, © iStockphoto.com / Florea Marius Catalin, © Arco Images GmbH / Alamy; p. 96 © Melissa Carroll , © sciencephotos / Alamy; p. 98 © iStockphoto.com / Laurence Gough, © Martyn F. Chillmaid / Science Photo Library; p. 100 © Lawrence Migdale / Science Photo Library; p. 102 © Martyn F. Chillmaid / Science Photo Library; p. 103 © iStockphoto.com / Joan Vicent Cantó Roig; p. 108 © Buzz Pictures / Alamy, © Buzz Pictures / Alamy, © Buzz Pictures / Alamy; p. 110 © iStockphoto.com / Sim Kay Seng, © James D. Watt / imagequest3d.com, © 2008 Jupiterimages Corporation; p. 111 © Arch White / Alamy, © Dave Ellison / Alamy; p. 112 © Royal Geographical Society, © Jim Edds / Science Photo Library, © 2004 Topham Picturepoint; p. 113 © 2008 Jupiterimages Corporation, © iStockphoto.com / Joshua Blake, © Jan Greune / Stock4B / Getty Images; p. 114 © Alexis Rosenfeld / Science Photo Library; p. 115 © iStockphoto.com / Trevor Fisher; p. 116 © The Stock Asylum, LLC / Alamy, © iStockphoto.com / Kativ, © iStockphoto.com / Jim Parkin, © iStockphoto.com / Trevor Fishe, © iStockphoto.com / Brian Sullivan, © Seb Rogers / Alamy; p. 119 © iStockphoto.com / Michael Svoboda, © General Photographic Agency / Hulton Archive / Getty Images, © Archivberlin Fotoagentur GmbH / Alamy; p. 121 © iStockphoto.com / Rob Cruse; p. 122 © Graham Harrison / Alamy; p. 123 © Popperfoto / Getty Images, © Phil Cole / Getty Images Sport, © Alan Cheek / Alamy, © iStockphoto.com / zennie, © Digital Vision / Alamy; p. 124 © Photofusion Picture Library / Alamy; p. 125 © Richard Wareham Fotografie / Alamy; p. 130 © Professor David Wallace and the MIT mechanical engineering class 2.009, Product Engineering Processes; p. 131 © Mary Evans Picture Library, © Adam Hart-Davis / Science Photo Library; p. 132 © iStockphoto.com / Juuce Interactive, © iStockphoto.com / Dan Barnes; p. 133 © Anthony Collins / Alamy, © allOver photography / Alamy, © iStockphoto.com / Karin Sass; p. 134 © Valery Sibrikov. Image from BigStockPhoto.com, © iStockphoto.com / Milena Sobieraj, © moodboard – Fotolia.com, © Kim Karpeles / Alamy, © Mark McEvoy / Alamy, © Ariel Skelley / Corbis, © Simon Clay / Alamy; p. 135 © Daniel H. Bailey / Alamy, © Catherina Holder. Image from BigStockPhoto.com, © iStockphoto.com / Tom Hahn, © iStockphoto.com / Dave Logan; p. 137 © David Page / Alamy, © Chris Brink. Image from BigStockPhoto.com, © iStockphoto.com / Clayton Hansen; p. 138 © iStockphoto.com / Tor Lindqvist, © iStockphoto.com / Mark Evans; p. 139 © iStockphoto.com / Sergey Kashkin, © Philippe Psaila / Science Photo Library; p. 140 © Erich Schrempp / Science Photo Library, © iStockphoto.com / Evgeny Terentyev; p. 141 © Robin Scagell / Science Photo Library; p. 142 © Sheila Terry / Science Photo Library, © Science Photo Library; p. 143 © David Parker / Science Photo Library; p. 144 © iStockphoto.com / John M. Chase, © iStockphoto.com / Terraxplorer; p. 146 © iStockphoto.com / Oleg Prikhodko, © iStockphoto.com / Gualtiero Boffi, © David Schleser / Nature's Images / Science Photo Library, © Georgette Douwma / Science Photo Library, © Dmitriy Chistoprudov. Image from BigStockPhoto.com; p. 147 © Straw Hat / Alamy, © iStockphoto.com / Oktay Ortakcioglu; p. 148 © Science Source / Science Photo Library, © Sinclair Stammers / Science Photo Library, © Susumu Nishinaga / Science Photo Library, © Susan & Allan Parker / Alamy; p. 149 © iStockphoto.com / Celso Diniz, © iStockphoto.com / Rui Saraiva, © iStockphoto.com / Alexander Hafemann, © iStockphoto.com / Eline Spek; p. 150 © CC Studio / Science Photo Library, © Ria Novosti / Science Photo Library, © Tek Image / Science Photo Library, © Will & Deni McIntyre / Science Photo Library, © Tek Image / Science Photo Library, © Mauro Fermariello / Science Photo Library; p. 152 © iStockphoto.com / Ovidiu Iordachi, © Photo Researchers / Science Photo Library, © 2008 Jupiterimages Corporation; p. 156 © iStockphoto.com / Gary Martin, © iStockphoto.com / dwphotos; p. 157 © Robert Brook / Science Photo Library, © Ian Hooton / Science Photo Library, © iStockphoto.com / czardases; p. 158 © iStockphoto.com / Igor Smichkov, © Robyn Mackenzie. Image from BigStockPhoto.com; p. 164 © Tim Davis / Corbis, © Ralph Lee Hopkins / Science Photo Library; p. 165 © Sinclair Stammers / Science Photo Library; p. 166 © Jacques Jangoux / Alamy; p. 167 © Charles O'Rear / Corbis; p. 168 © Alfred Pasieka / Science Photo Library, © Alfred Pasieka / Science Photo Library, © Alfred Pasieka / Science Photo Library, © FLPA / Alamy; p. 169 © Veronique Leplat / Science Photo Library; p. 170 © Robert Brook / Science Photo Library, © Peter Ryan / Science Photo Library, © 1994 A. Palotas, A. Sarofim & J. Vander Sande / Science Photo Library; p. 171 © Stacey Brown. Image from BigStockPhoto.com; p. 172 © iStockphoto.com / Sava Miokovic, © iStockphoto.com / jacus, © iStockphoto.com / Manuela Dabkowska, © iStockphoto.com / Aleksandr Lukin; p. 174 © iStockphoto.com / David Parsons, © Jeremy Walker / Science Photo Library; p. 175 © NASA, coloured by John Wells / Science Photo Library; p. 176 © Visions of America, LLC / Alamy, © Gerolf Kalt / zefa / Corbis, © Stephen Finn. Image from BigStockPhoto.com, © Simon Fraser / Science Photo Library; p. 177 © Nigel Cattlin / Alamy, © Alexey Lisovoy. Image from BigStockPhoto.com, © Martin Heaney. Image from BigStockPhoto.com, © Nigel Cattlin / Alamy, © Kris Langston. Image from BigStockPhoto.com, © Agnes Martelet. Image from BigStockPhoto.com; p. 178 © iStockphoto.com / Ruth Black, © iStockphoto.com / John Anderson, © iStockphoto.com / Michael Flippo, © iStockphoto.com / Adrian Baddeley, © iStockphoto.com / Marcus Horn, © iStockphoto.com / Richard Moon; p. 179 © Frans Lanting / Corbis, © iStockphoto.com / Ryan Jones; p. 180 © iStockphoto.com / Tony Tremblay; p. 181 © iStockphoto.com / Daniel Kourey, © iStockphoto.com / rememp. 186 © Danilo Ascione. Image from BigStockPhoto.com; p. 186 © iStockphoto.com / Soubrette; p. 187 © Pasquale Sorrentino / Science Photo Library; p. 188 © David Deveney. Image from BigStockPhoto.com; p. 189 © iStockphoto.com / mikeuk, © Robert Harding Picture Library Ltd / Alamy, © Laurance Richardson / Alamy; p. 190 © Dirk Wiersma / Science Photo Library, © Peter Mautsch. Image from BigStockPhoto.com, © iStockphoto.com / Ling Xia, © John Warburton-Lee Photography / Alamy; p. 191 © Pascal Goetgheluck / Science Photo Library; p. 192 © The Natural History Museum / Alamy; p. 194 © iStockphoto.com / Black Beck Photographic, © iStockphoto.com / Stuart Murchison; p. 195 © iStockphoto.com / mikeuk; p. 196 © Lera Titova. Image from BigStockPhoto.com; p. 197 © Jens Mayer. Image from BigStockPhoto.com, © Stephen Bisgrove / Alamy, © iStockphoto.com / Andrew Hyslop, © Dirk Wiersma / Science Photo Library; p. 198 © Andrew Lambert Photography / Science Photo Library, © iStockphoto.com / mikeuk, © 2008 Jupiterimages Corporation; p. 200 © Arctic Images / Alamy; p. 201 © Mark A. Schneider / Science Photo Library; p. 201 © iStockphoto.com / saints4757; p. 203 © NASA / Science Photo Library, © iStockphoto.com / Martin Bowker